Studies in Computational Intelligence

Volume 686

Series editor

Janusz Kacprzyk, Polish Academy of Sciences, Warsaw, Poland
e-mail: kacprzyk@ibspan.waw.pl

About this Series

The series "Studies in Computational Intelligence" (SCI) publishes new developments and advances in the various areas of computational intelligence—quickly and with a high quality. The intent is to cover the theory, applications, and design methods of computational intelligence, as embedded in the fields of engineering, computer science, physics and life sciences, as well as the methodologies behind them. The series contains monographs, lecture notes and edited volumes in computational intelligence spanning the areas of neural networks, connectionist systems, genetic algorithms, evolutionary computation, artificial intelligence, cellular automata, self-organizing systems, soft computing, fuzzy systems, and hybrid intelligent systems. Of particular value to both the contributors and the readership are the short publication timeframe and the worldwide distribution, which enable both wide and rapid dissemination of research output.

More information about this series at http://www.springer.com/series/7092

Erik Cuevas · Valentín Osuna
Diego Oliva

Evolutionary Computation Techniques: A Comparative Perspective

 Springer

Erik Cuevas
CUCEI
Universidad de Guadalajara
Guadalajara
Mexico

Diego Oliva
Tecnológico de Monterrey
Campus Guadalajara
Guadalajara
Mexico

Valentín Osuna
CUTONALÁ
Universidad de Guadalajara
Tonalá Jalisco
Mexico

and

CUCEI
Universidad de Guadalajara
Guadalajara
Mexico

ISSN 1860-949X   ISSN 1860-9503   (electronic)
Studies in Computational Intelligence
ISBN 978-3-319-84568-5   ISBN 978-3-319-51109-2   (eBook)
DOI 10.1007/978-3-319-51109-2

Printed on acid-free paper

This Springer imprint is published by Springer Nature
The registered company is Springer International Publishing AG
The registered company address is: Gewerbestrasse 11, 6330 Cham, Switzerland

Preface

Many problems in engineering nowadays concern with the goal of an "optimal" solution. Several optimization methods have therefore emerged, being researched and applied extensively to different optimization problems.

Typically, optimization methods arising in engineering are computationally complex because they require evaluation of a quite complicated objective function which is often multimodal, non-smooth or even discontinuous. The difficulties associated with using mathematical optimization on complex engineering problems have contributed to the development of alternative solutions. Evolutionary computation (EC) techniques are stochastic optimization methods that have been developed to obtain near-optimum solutions in complex optimization problems, for which traditional mathematical techniques normally fail.

EC methods use as inspiration our scientific understanding of biological, natural or social systems, which at some level of abstraction can be represented as optimization processes. In their operation, searcher agents emulate a group of biological or social entities which interact with each other based on specialized operators that model a determined biological or social behavior. These operators are applied to a population (or several sub-populations) of candidate solutions (individuals) that are evaluated with respect to their fitness. Thus in the evolutionary process, individual positions are successively approximated to the optimal solution of the system to be solved.

Due to their robustness, EC techniques are well-suited options for industrial and real-world tasks. They do not need gradient information and they can operate on each kind of parameter space (continuous, discrete, combinatorial, or even mixed variants). Essentially, the credibility of evolutionary algorithms relies on their ability to solve difficult real-world problems with the minimal amount of human effort.

There exist some common features clearly appear in most of the EC approaches, such as the use of diversification to force the exploration of regions of the search space, rarely visited until now, and the use of intensification or exploitation, to investigate thoroughly some promising regions. Another common feature is the use of memory to archive the best solutions encountered.

EC techniques are used to estimate the solutions to complex optimization problems. They are often designed to meet the requirements of particular problems because no single optimization algorithm can solve all problems competitively. Therefore, in order to select an appropriate EC technique, its relative efficacy must be appropriately evaluated.

Several comparisons among ECT have been reported in the literature. Nevertheless, they suffer from one limitation: their conclusions are based on the performance of popular evolutionary approaches over a set of synthetic functions with exact solutions and well-known behaviors, without considering the application context or including recent developments.

Numerous books have been published taking in account many of the most widely known methods, namely simulated annealing, tabu search, evolutionary algorithms, ant colony algorithms, particle swarm optimization or differential evolution, but attempts to consider the discussion of alternative approaches are scarce. The excessive publication of developments based on the simple modification of popular EC methods present an important disadvantage, in that it distracts attention away from other innovative ideas in the field of EC. There exist several alternative EC methods which consider very interesting concepts; however, they seem to have been completely overlooked in favor of the idea of modifying, hybridizing or restructuring traditional EC approaches.

The goal of this book is to present the comparison of various EC techniques when they face complex optimization problems extracted from different engineering domains. In the comparisons, special attention is paid to recently developed algorithms. This book has been structured so that each chapter can be read first independently from the others. In each chapter, a complex engineering optimization problem is posed. Then, a particular EC technique is presented as the best choice, according to its search characteristics. Finally, a set of experiments are conducted in order to compare its performance to other popular EC methods.

Chapter 1 describes evolutionary computation (EC). This chapter concentrates on elementary concepts of evolutionary algorithms. Readers that are familiar with EC could skip this chapter.

In Chap. 2, the problem of multilevel segmentation in images is presented. In the approach, the Electromagnetism-Like algorithm (EMO) algorithm is proposed as the best option to find the optimal threshold values by maximizing the Tsallis entropy. In the approach, the algorithm uses as particles the encoding of a set of candidate threshold points. An objective function evaluates the segmentation quality of the candidate threshold points. Guided by the values of this objective function, the set of encoded candidate solutions are modified by using the EMO operators so that they can improve their segmentation quality as the optimization process evolves. The approach is compared to the cuckoo search algorithm (CSA) and the particle swarm optimization (PSO).

In Chap. 3, the problem of detecting circular shapes from complicated and noisy images is considered. The detection process is approached as a multimodal optimization problem. The artificial bee colony (ABC) algorithm is presented as the best possibility to solve the recognition problem. In the method, the ABC algorithm searches the entire edge-map looking for circular shapes by using the combination of three non-collinear edge points that represent candidate circles (food source locations) in the edge-only image of the scene. An objective function is used to measure the existence of a candidate circle over the edge map. Guided by the values of such objective function, the set of encoded candidate circles are evolved through the ABC algorithm so that the best candidate can be fitted into the most circular shape within the edge-only image. A subsequent analysis of the incorporated exhausted-source memory is then executed in order to identify potential useful local optima (other circles). The approach generates a fast sub-pixel detector which can effectively identify multiple circles in real images despite circular objects exhibiting significant occluded sections. Experimental evidence shows the effectiveness of the method for detecting circles under various conditions. A comparison to one state-of-the-art genetic algorithm-based method and the bacterial foraging optimization algorithm (BFOA) on different images has been included to demonstrate the performance of the proposed approach. Conclusions of the experimental comparison are validated through statistical tests that properly support the discussion.

In Chap. 4, the application of template matching (TM) is considered. TM plays an important role in several image processing applications. In a TM approach, it is sought the point in which it is presented the best possible resemblance between a sub-image known as template and its coincident region within a source image. TM involves two critical aspects: similarity measurement and search strategy. The simplest available TM method finds the best possible coincidence between the images through an exhaustive computation of the normalized cross-correlation (NCC) values (similarity measurement) for all elements of the source image (search strategy). In the chapter, the social spider optimization (SSO) algorithm is proposed to reduce the number of search locations in the TM process. The SSO algorithm is based on the simulation of cooperative behavior of social spiders. The algorithm considers two different search individuals (spiders): males and females. Depending on gender, each individual is conducted by a set of different evolutionary operators which mimic different cooperative behaviors that are typically found in the colony. In the proposed approach, spiders represent search locations which move throughout the positions of the source image. The NCC coefficient, used as a fitness value, evaluates the matching quality presented between the template image and the coincident region of the source image, for a determined search position (spider). The number of NCC evaluations is reduced by considering a memory which stores the NCC values previously visited in order to avoid the re-evaluation of the same search locations. Guided by the fitness values (NCC coefficients), the set of encoded candidate positions are evolved using the SSO operators until the best possible resemblance has been found. The approach is compared to imperialist competitive algorithm (ICA) and the particle swarm optimization (PSO).

In Chap. 5, the problem of motion estimation is presented. Motion estimation is a major problem for video-coding applications. Among several other motion estimation approaches, block matching (BM) algorithms are the most popular methods due to their effectiveness and simplicity at their software and hardware implementation. The BM approach assumes that the pixel movement inside a given region of the current frame (Macro-Block, MB) can be modeled as a pixel translation from its corresponding region in the previous frame. In this procedure, the motion vector is obtained by minimizing the sum of absolute differences (SAD) from the current frame's MB over a determined search window from the previous frame. Unfortunately, the SAD evaluation is computationally expensive and represents the most time-consuming operation in the BM process. The simplest available BM method is the full search algorithm (FSA) which finds the most accurate motion vector through an exhaustive computation of SAD values for all elements of the search window. However, several fast BM algorithms have been lately proposed to reduce the number of SAD operations by calculating only a fixed subset of search locations at the price of poor accuracy. In this chapter, the differential evolution (DE) method is proposed to reduce the number of search locations in the BM process. In order to avoid the computing of several search locations, the algorithm estimates the SAD (fitness) values for some locations by considering SAD values from previously calculated neighboring positions. The approach is compared to the popular particle swarm optimization (PSO).

In Chap. 6, the application of modeling solar cells is presented. In order to improve the performance of solar energy systems, accurate modeling of current versus voltage (I–V) characteristics of solar cells has attracted the attention of various researches. The main drawback in accurate modeling is the lack of information about the precise parameter values which indeed characterize the solar cell. Since such parameters cannot be extracted from the datasheet specifications, an optimization technique is necessary to adjust experimental data to the solar cell model. Considering the I–V characteristics of solar cells, the optimization task involves the solution of complex nonlinear and multimodal objective functions. Several optimization approaches have been proposed to identify the parameters of solar cells. However, most of them obtain suboptimal solutions due to their premature convergence and their difficulty to overcome local minima in multimodal problems. This chapter proposes the use of the artificial bee colony (ABC) algorithm to accurately identify the solar cells' parameters. In comparison with other evolutionary algorithms, ABC exhibits a better search capacity to face multimodal objective functions. In order to illustrate the proficiency of the proposed approach, it is compared to other well-known optimization methods. Experimental results demonstrate the high performance of the proposed method in terms of robustness and accuracy.

Chapter 7 presents the problem of parameter identification in induction motors. Induction motors represent the main component in most of the industries. They consume the highest energy percentages in industrial facilities. This energy consumption depends on the operation conditions of the induction motor imposed by its internal parameters. Since the internal parameters of an induction motor are not directly measurable, an identification process must be conducted to obtain them. In the identification process, the parameter estimation is transformed into a multidimensional optimization problem where the internal parameters of the induction motor are considered as decision variables. Under this approach, the complexity of the optimization problem tends to produce multimodal error surfaces for which their cost functions are significantly difficult to minimize. This chapter presents an algorithm for the optimal parameter identification of induction motors. To determine the parameters, the proposed method uses a recent evolutionary method called the gravitational search algorithm (GSA). Different from most of existent evolutionary algorithms, GSA presents a better performance in multimodal problems, avoiding critical flaws such as the premature convergence to suboptimal solutions. The approach is compared to the popular particle swarm optimization (PSO), differential evolution (DE) and artificial bee colony (ABC).

In Chap. 8, the application of white blood cells (WBC) detection in images is presented. The automatic detection of WBC still remains an unsolved issue in medical imaging. The analysis of WBC images has engaged researchers from fields of medicine and computer vision alike. Since WBC can be approximated by an ellipsoid form, an ellipse detector algorithm may be successfully applied in order to recognize them. In this chapter, the differential evolution (DE) algorithm is used for the automatic detection of WBC embedded into complicated and cluttered smear images. The approach transforms the detection task into an optimization problem where individuals emulate candidate ellipses. An objective function evaluates if such candidate ellipses are really present in the edge image of the smear. Guided by the values of such function, the set of encoded candidate ellipses (individuals) are evolved using the DE algorithm so that they can fit into the WBC enclosed within the edge-only map of the image. Experimental results from white blood cell images with a varying range of complexity are included to validate the efficiency of the proposed technique in terms of accuracy and robustness.

Chapter 9 presents the problem of estimating view transformations from image correspondences. Many computer vision algorithms include a robust estimation step where model parameters are computed from a dataset containing a significant proportion of outliers. Based on different criteria, several robust techniques have been suggested to solve such a problem, being the random sampling consensus (RANSAC) algorithm the most well-known. In this chapter, a method for robustly estimating multiple view relations from point correspondences is posed. The approach combines the RANSAC method and the harmony search (HS) algorithm. With the combination, the proposed method adopts a different sampling strategy than RANSAC to generate putative solutions. Under the new mechanism, at each iteration, new candidate solutions are built taking into account the quality of the models generated by previous candidate solutions, rather than purely random as is the case of RANSAC. The rules for the generation of candidate solutions (samples) are motivated by the improvisation process that occurs when a musician searches for a better state of harmony. As a result, the proposed approach can substantially reduce the number of iterations still preserving the robust capabilities of RANSAC. The method is generic and its use is illustrated by the estimation of homographies, considering synthetic and real images. The approach is compared to the popular particle swarm optimization (PSO).

Finally, in Chap. 10, the problem of identification of infinite impulse response (IIR) models is posed. System identification is a complex optimization problem which has recently attracted the attention in the field of science and engineering.

In particular, the use of infinite impulse response (IIR) models for identification is preferred over their equivalent finite impulse response (FIR) models since the former yield more accurate models of physical plants for real-world applications. However, IIR structures tend to produce multimodal error surfaces for which their cost functions are significantly difficult to minimize. This chapter presents the comparison of various evolutionary computation optimization techniques applied to IIR model identification. In the comparison, special attention is paid to recently developed algorithms such as cuckoo search and flower pollination algorithm, also including popular approaches. Results over several models are presented and statistically validated.

The material has been compiled from a teaching perspective. For this reason, the book is primarily intended for undergraduate and postgraduate students of science, engineering, or computational mathematics. It can be appropriate for courses such as Artificial Intelligence, Evolutionary Computation, Computational Intelligence, etc. Likewise, the material can be useful for researches from the evolutionary computation and artificial intelligence communities. The important purpose of this book is to bridge the gap between evolutionary optimization techniques and complex engineering applications. Therefore, researchers, who are familiar with popular evolutionary computation approaches, will appreciate that the techniques discussed are beyond simple optimization tools since they have been adapted to solve significant problems that commonly arise on several engineering domains. On the other hand, students of the evolutionary computation community can prospect new research niches for their future work as master or Ph.D. thesis.

As authors, we wish to thank many people who were somehow involved in the writing process of this book. We thank Dr. Carlos Coello Coello and Dr. Raul Rojas for supporting us to have it published. We express our gratitude to Prof. Janusz kacprzyk, who so warmly sustained this project. Acknowledgements also go to Dr. Thomas Ditzinger, who so kindly agreed to its appearance. We also acknowledge that this work was supported by CONACYT under the grant number CB 181053.

Guadalajara, Mexico
November 2016

Erik Cuevas
Valentín Osuna
Diego Oliva

Contents

Chapter 1
Introduction

Abstract The objective of this chapter is to motivate the use of evolutionary techniques for solving optimization problems. The chapter is conducted in such a way that it is clear the necessity of using evolutionary optimization methods for the solution of complex problems present in engineering. The chapter also gives an introduction to the optimization techniques, considering their main characteristics.

1.1 Definition of an Optimization Problem

The vast majority of engineering algorithms use some form of optimization, as they intend to find some solution which is "best" according to some criterion. From a general perspective, an optimization problem is a situation that requires to decide for a choice from a set of possible alternatives in order to reach a predefined/ required benefit at minimal costs [1].

Consider a public transportation system of a city, for example. Here the system has to find the "best" route to a destination location. In order to rate alternative solutions and eventually find out which solution is "best," a suitable criterion has to be applied. A reasonable criterion could be the distance of the routes. We then would expect the optimization algorithm to select the route of shortest distance as a solution. Observe, however, that other criteria are possible, which might lead to different "optimal" solutions, e.g., number of transfers, ticket price or the time it takes to travel the route leading to the fastest route as a solution.

Mathematically speaking, optimization can be described as follows: Given a function $f : S \rightarrow \mathbb{R}$ which is called the objective function, find the argument which minimizes f:

$$x^* = \arg\min_{x \in S} f(x) \tag{1.1}$$

S defines the so-called solution set, which is the set of all possible solutions for the optimization problem. Sometimes, the unknown(s) x are referred to design

E. Cuevas et al., *Evolutionary Computation Techniques:*
A Comparative Perspective, Studies in Computational Intelligence 686,
DOI 10.1007/978-3-319-51109-2_1

variables. The function f describes the optimization criterion, i.e., enables us to calculate a quantity which indicates the "quality" of a particular x.

In our example, S is composed by the subway trajectories and bus lines, etc., stored in the database of the system, x is the route the system has to find, and the optimization criterion $f(x)$ (which measures the quality of a possible solution) could calculate the ticket price or distance to the destination (or a combination of both), depending on our preferences.

Sometimes there also exist one or more additional constraints which the solution x^* has to satisfy. In that case we talk about constrained optimization (opposed to unconstrained optimization if no such constraint exists). As a summary, an optimization problem has the following components:

- One or more design variables x for which a solution has to be found
- An objective function $f(x)$ describing the optimization criterion
- A solution set S specifying the set of possible solutions x
- (optional) One or more constraints on x.

In order to be of practical use, an optimization algorithm has to find a solution in a reasonable amount of time with reasonable accuracy. Apart from the performance of the algorithm employed, this also depends on the problem at hand itself. If we can hope for a numerical solution, we say that the problem is well-posed. For assessing whether an optimization problem is well-posed, the following conditions must be fulfilled:

1. A solution exists.
2. There is only one solution to the problem, i.e., the solution is unique.
3. The relationship between the solution and the initial conditions is such that small perturbations of the initial conditions result in only small variations of x^*.

1.2   Classical Optimization

Once a task has been transformed into an objective function minimization problem, the next step is to choose an appropriate optimizer. Optimization algorithms can be divided in two groups: derivative-based and derivative-free [2].

In general, $f(x)$ may have a nonlinear form respect to the adjustable parameter x. Due to the complexity of $f(\cdot)$, in classical methods, it is often used an iterative algorithm to explore the input space effectively. In iterative descent methods, the next point x_{k+1} is determined by a step down from the current point x_k in a direction vector $\mathbf{d}$:

$$x_{k+1} = x_k + \alpha \mathbf{d}, \tag{1.2}$$

where α is a positive step size regulating to what extent to proceed in that direction. When the direction d in Eq. 1.1 is determined on the basis of the gradient (**g**) of the objective function $f(\cdot)$, such methods are known as gradient-based techniques.

The method of steepest descent is one of the oldest techniques for optimizing a given function. This technique represents the basis for many derivative-based methods. Under such a method, the Eq. 1.3 becomes the well-known gradient formula:

$$x_{k+1} = x_k - \alpha \mathbf{g}(f(x)), \tag{1.3}$$

However, classical derivative-based optimization can be effective as long the objective function fulfills two requirements:

- The objective function must be two-times differentiable.
- The objective function must be uni-modal, i.e., have a single minimum.

A simple example of a differentiable and uni-modal objective function is

$$f(x_1, x_2) = 10 - e^{-\left(x_1^2 + 3 \cdot x_2^2\right)} \tag{1.4}$$

Figure 1.1 shows the function defined in Eq. 1.4.

Unfortunately, under such circumstances, classical methods are only applicable for a few types of optimization problems. For combinatorial optimization, there is no definition of differentiation.

Furthermore, there are many reasons why an objective function might not be differentiable. For example, the "floor" operation in Eq. 1.5 quantizes the function in Eq. 1.4, transforming Fig. 1.1 into the stepped shape seen in Fig. 1.2. At each step's edge, the objective function is non-differentiable:

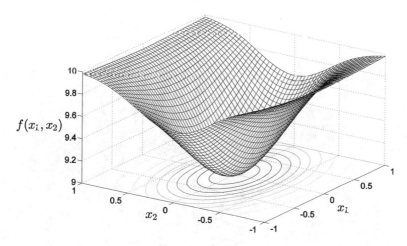

Fig. 1.1 Uni-modal objective function

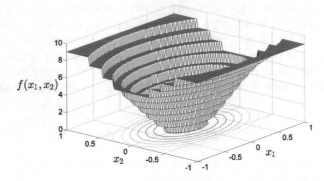

Fig. 1.2 A non-differentiable, quantized, uni-modal function

$$f(x_1, x_2) = \text{floor}\left(10 - e^{-\left(x_1^2 + 3 \cdot x_2^2\right)}\right) \tag{1.5}$$

Even in differentiable objective functions, gradient-based methods might not work. Let us consider the minimization of the Griewank function as an example.

$$\begin{aligned} \text{minimize} \quad & f(x_1, x_2) = \frac{x_1^2 + x_2^2}{4000} - \cos(x_1)\cos\left(\frac{x_2}{\sqrt{2}}\right) + 1 \\ \text{subject to} \quad & -30 \le x_1 \le 30 \\ & -30 \le x_2 \le 30 \end{aligned} \tag{1.6}$$

From the optimization problem formulated in Eq. 1.6, it is quite easy to understand that the global optimal solution is $x_1 = x_2 = 0$. Figure 1.3 visualizes the function defined in Eq. 1.6. According to Fig. 1.3, the objective function has many local optimal solutions (multimodal) so that the gradient methods with a randomly generated initial solution will converge to one of them with a large probability.

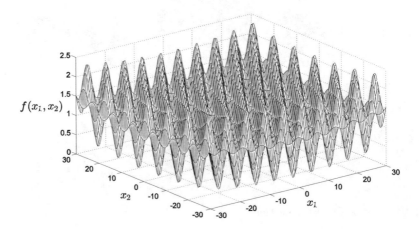

Fig. 1.3 The Griewank multi-modal function

Considering the limitations of gradient-based methods, engineering problems make difficult their integration with classical optimization methods. Instead, some other techniques which do not make assumptions and which can be applied to wide range of problems are required [3].

1.3 Evolutionary Computation Methods

Evolutionary computation (EC) [4] methods are derivative-free procedures, which do not require that the objective function must be neither two-times differentiable nor uni-modal. Therefore, EC methods as global optimization algorithms can deal with non-convex, nonlinear, and multimodal problems subject to linear or nonlinear constraints with continuous or discrete decision variables.

The field of EC has a rich history. With the development of computational devices and demands of industrial processes, the necessity to solve some optimization problems arose despite the fact that there was not sufficient prior knowledge (hypotheses) on the optimization problem for the application of a classical method. In fact, in the majority of engineering applications, the problems are highly nonlinear, or characterized by a noisy fitness, or without an explicit analytical expression as the objective function might be the result of an experimental or simulation process. In this context, the EC methods have been proposed as optimization alternatives.

An EC technique is a general method for solving optimization problems. It uses an objective function in an abstract and efficient manner, typically without utilizing deeper insights into its mathematical properties. EC methods do not require hypotheses on the optimization problem nor any kind of prior knowledge on the objective function. The treatment of objective functions as "black boxes" [5] is the most prominent and attractive feature of EC methods. EC methods obtain knowledge about the structure of an optimization problem by utilizing information obtained from the possible solutions (i.e., candidate solutions) evaluated in the past. This knowledge is used to construct new candidate solutions which are likely to have a better quality.

Recently, several EC methods have been proposed with interesting results. Such approaches uses as inspiration our scientific understanding of biological, natural or social systems, which at some level of abstraction can be represented as optimization processes [6]. These methods include the social behavior of bird flocking and fish schooling such as the Particle Swarm Optimization (PSO) algorithm [7], the cooperative behavior of bee colonies such as the Artificial Bee Colony (ABC) technique [8], the improvisation process that occurs when a musician searches for a better state of harmony such as the Harmony Search (HS) [9], the emulation of the bat behavior such as the Bat Algorithm (BA) method [10], the mating behavior of firefly insects such as the Firefly (FF) method [11], the social-spider behavior such as the Social Spider Optimization (SSO) [12], the simulation of the animal behavior in a group such as the Collective Animal

Behavior [13], the emulation of immunological systems as the clonal selection algorithm (CSA) [14], the simulation of the electromagnetism phenomenon as the electromagnetism-Like algorithm [15], and the emulation of the differential and conventional evolution in species such as the Differential Evolution (DE) [16] and Genetic Algorithms (GA) [17], respectively. Their effective search strategies have motivated their use to solve an extensive variety of engineering applications such as image processing [18–20], energy [21] and signal processing [22].

1.3.1   Structure of an Evolutionary Computation Algorithm

From a conventional point of view, an EC method is an algorithm that simulates at some level of abstraction a biological, natural or social system. To be more specific, a standard EC algorithm includes:

1. One or more populations of candidate solutions are considered.
2. These populations change dynamically due to the production of new solutions.
3. A fitness function reflects the ability of a solution to survive and reproduce.
4. Several operators are employed in order to explore an exploit appropriately the space of solutions.

The EC methodology suggest that, on average, candidate solutions improve their fitness over generations (i.e., their capability of solving the optimization problem). A simulation of the evolution process based on a set of candidate solutions whose fitness is properly correlated to the objective function to optimize will, on average, lead to an improvement of their fitness and thus steer the simulated population towards the global solution.

Most of the optimization methods have been designed to solve the problem of finding a global solution of a nonlinear optimization problem with box constraints in the following form:

$$\begin{array}{ll} \text{maximize} & f(\mathbf{x}), \quad \mathbf{x} = (x_1, \ldots, x_d) \in \mathbb{R}^d \\ \text{subject to} & \mathbf{x} \in \mathbf{X} \end{array} \qquad (1.7)$$

where $f : \mathbb{R}^d \to \mathbb{R}$ is a nonlinear function whereas $\mathbf{X} = \{\mathbf{x} \in \mathbb{R}^d | l_i \leq x_i \leq u_i, \quad i = 1, \ldots, d.\}$ is a bounded feasible search space, constrained by the lower (l_i) and upper (u_i) limits.

In order to solve the problem formulated in Eq. 1.6, in an evolutionary computation method, a population $\mathbf{P}^k$ ($\{\mathbf{p}_1^k, \mathbf{p}_2^k, \ldots, \mathbf{p}_N^k\}$) of N candidate solutions (individuals) evolves from the initial point $(k = 0)$ to a total *gen* number iterations $(k = gen)$. In its initial point, the algorithm begins by initializing the set of N candidate solutions with values that are randomly and uniformly distributed between the pre-specified lower (l_i) and upper (u_i) limits. In each iteration, a set of evolutionary operators are applied over the population $\mathbf{P}^k$ to build the new

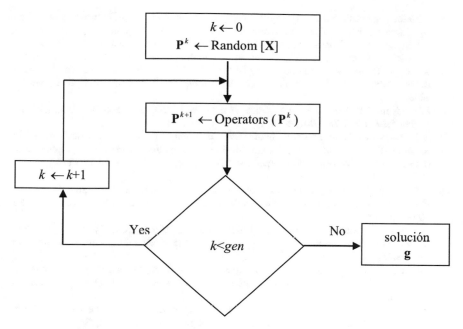

Fig. 1.4 The basic cycle of a EC method

population $\mathbf{P}^{k+1}$. Each candidate solution $\mathbf{p}_i^k$ ($i \in [1,\ldots,N]$) represents a d-dimensional vector $\left\{p_{i,1}^k, p_{i,2}^k, \ldots, p_{i,d}^k\right\}$ where each dimension corresponds to a decision variable of the optimization problem at hand. The quality of each candidate solution $\mathbf{p}_i^k$ is evaluated by using an objective function $f\left(\mathbf{p}_i^k\right)$ whose final result represents the fitness value of $\mathbf{p}_i^k$. During the evolution process, the best candidate solution $\mathbf{g}$ ($g_1, g_2, \ldots g_d$) seen so-far is preserved considering that it represents the best available solution. Figure 1.4 presents a graphical representation of a basic cycle of a EC method.

References

1. Bahriye Akay, Dervis Karaboga, A survey on the applications of artificial bee colony in signal, image, and video processing, Signal, Image and Video Processing, 9(4), (2015), 967–990.
2. Xin-She Yang, Engineering Optimization, 2010, John Wiley & Sons, Inc.
3. Marco Alexander Treiber, Optimization for Computer Vision An Introduction to Core Concepts and Methods, Springer, 2013.
4. Dan Simon, Evolutionary Optimization Algorithms, Wiley, 2013.
5. Blum, C., Roli, A.: Metaheuristics in Combinatorial Optimization: Overview and Conceptual Comparison. ACM Computing Surveys (CSUR) 35(3), 268–308 (2003); doi:10.1145/937503.937505.

6. Satyasai Jagannath Nanda, Ganapati Panda, A survey on nature inspired metaheuristic algorithms for partitional clustering, *Swarm and Evolutionary Computation*, 16, (2014), 1–18.
7. J. Kennedy and R. Eberhart, Particle swarm optimization, in Proceedings of the 1995 IEEE International Conference on Neural Networks, vol. 4, pp. 1942–1948, December 1995.
8. Karaboga, D. An Idea Based on Honey Bee Swarm for Numerical Optimization. Technical Report-TR06. Engineering Faculty, Computer Engineering Department, Erciyes University, 2005.
9. Z.W. Geem, J.H. Kim, G.V. Loganathan, A new heuristic optimization algorithm: harmony search, *Simulations* 76 (2001) 60–68.
10. X.S. Yang, A new metaheuristic bat-inspired algorithm, in: C. Cruz, J. González, G.T.N. Krasnogor, D.A. Pelta (Eds.), Nature Inspired Cooperative Strategies for Optimization (NISCO 2010), Studies in Computational Intelligence, vol. 284, Springer Verlag, Berlin, 2010, pp. 65–74.
11. X.S. Yang, Firefly algorithms for multimodal optimization, in: Stochastic Algorithms: Foundations and Applications, SAGA 2009, Lecture Notes in Computer Sciences, vol. 5792, 2009, pp. 169–178.
12. Erik Cuevas, Miguel Cienfuegos, Daniel Zaldívar, Marco Pérez-Cisneros, A swarm optimization algorithm inspired in the behavior of the social-spider, *Expert Systems with Applications*, 40(16), (2013), 6374-6384.
13. Cuevas, E., González, M., Zaldivar, D., Pérez-Cisneros, M., García, G. An algorithm for global optimization inspired by collective animal behaviour, *Discrete Dynamics in Nature and Society* 2012, art. no. 638275.
14. L.N. de Castro, F.J. von Zuben, Learning and optimization using the clonal selection principle, IEEE Transactions on Evolutionary Computation 6 (3) (2002) 239–251.
15. Ş. I. Birbil and S. C. Fang, "An electromagnetism-like mechanism for global optimization," J. Glob. Optim., vol. 25, no. 1, pp. 263–282, 2003.
16. Storn, R., Price, K., 1995. Differential Evolution -a simple and efficient adaptive scheme for global optimisation over continuous spaces. Technical ReportTR-95–012, ICSI, Berkeley, CA.
17. D.E. Goldberg, Genetic Algorithm in Search Optimization and Machine Learning, Addison-Wesley, 1989.
18. Cuevas, E., Zaldivar, D., Pérez-Cisneros, M., Ramírez-Ortegón, M., Circle detection using discrete differential evolution Optimization, Pattern Analysis and Applications, 14 (1), (2011), 93–107.
19. Cuevas, E., Ortega-Sánchez, N., Zaldivar, D., Pérez-Cisneros, M., Circle detection by Harmony Search Optimization, Journal of Intelligent and Robotic Systems: Theory and Applications, 66(3), (2012), 359–376.
20. Oliva, D., Cuevas, E., Pajares, G., Zaldivar, D., Perez-Cisneros, M., Multilevel thresholding segmentation based on harmony search optimization, Journal of Applied Mathematics, 2013, 575414.
21. Oliva, D., Cuevas, E., Pajares, G., Parameter identification of solar cells using artificial bee colony optimization, Energy, 72, (2014), 93–102.
22. Cuevas, E., Gálvez, J., Hinojosa, S., Zaldívar, D., Pérez-Cisneros, M., A comparison of evolutionary computation techniques for IIR model identification, Journal of Applied Mathematics, 2014, 827206.

Chapter 2
Multilevel Segmentation in Digital Images

Abstract Segmentation is used to divide an image into separate regions, which in fact correspond to different real-world objects. One interesting functional criterion for segmentation is the Tsallis entropy (TE), which gives excellent results in bi-level thresholding. However, when it is applied to multilevel thresholding (MT), its evaluation becomes computationally expensive, since each threshold point adds restrictions, multimodality and complexity to its functional formulation. In this chapter, a new algorithm for multilevel segmentation based on the Electromagnetism-Like algorithm (EMO) is presented. In the approach, the EMO algorithm is used to find the optimal threshold values by maximizing the Tsallis entropy. Experimental results over several images demonstrate that the proposed approach is able to improve the convergence velocity, compared with similar methods such as Cuckoo search, and Particle Swarm Optimization.

2.1 Introduction

Segmentation is one of the basic steps of an image analysis system, and consists in separating objects from each other, by considering characteristics contained in a digital image [1]. It has been applied to feature extraction [2], object identification and classification [3], surveillance [4], among other areas. In order to obtain homogeneous regions of pixels, a common method is using the histogram's information with a thresholding approach [5]. This method is considered the easiest one in segmentation, and it works taking threshold values which separate adequately the distinct regions of pixels in the image being processed. In general, there are two thresholding approaches, namely bi-level and multilevel. In bi-level thresholding (BT), it is only needed a threshold value to separate the two objects of an image (e.g. foreground and background). For real life images, BT doesn't provide appropriate results. On the other hand, multilevel thresholding (MT) divides the pixels in more than two homogeneous classes and it needs several

© Springer International Publishing AG 2017
E. Cuevas et al., *Evolutionary Computation Techniques:
A Comparative Perspective*, Studies in Computational Intelligence 686,
DOI 10.1007/978-3-319-51109-2_2

threshold values [5, 6]. Threshold methods are divided in parametric and non-parametric [6, 7]. In parametric approaches, it is necessary estimating the parameters of a probability density function capable of modelling each class. Such an approach is time consuming and computationally expensive. A nonparametric technique employs a given criteria (between-class variance, entropy and error rate [9–8]) which must be optimized to determine the optimal threshold values. These approaches result an attractive option due their robustness and accuracy [9].

For bi-level thresholding there exist two classical methods: the first one, proposed by Otsu [10], maximizes the variance between classes, whereas the second one, proposed by Kapur in [11], uses the entropy maximization to measure the homogeneity among classes. Their efficiency and accuracy have been already proved by segmenting pixels into two classes [12]. Both methods, Otsu and Kapur, can be expanded for multilevel thresholding; however, their computational complexity is increased, and also its accuracy decreases with each new threshold added into the searching process [12, 13].

The Tsallis entropy (TE), proposed in [14], is known as the non-extensive entropy, and can be considered as an extension of Shannon's entropy. Recently, there exist several studies that report similarities among the Tsallis, the Shannon and the Boltzmann-Gibbs entropies [13, 16–15]. Different to the Otsu and Kapur methods, the Tsallis entropy produces a functional formulation whose accuracy does not depend on the number of threshold points [16]. In the process of image segmentation, under the TE perspective, it is selected a set of threshold values that maximize the TE functional formulation, so that each pixel is assigned to a determined class according to its corresponding threshold points. TE gives excellent results in bi-level thresholding. However, when it is applied to multilevel thresholding (MT), its evaluation becomes computationally expensive, since each threshold point adds restrictions, multimodality and complexity to its functional formulation. Therefore, in the process of finding the appropriate threshold values, it is desired to limit the number of evaluations of the TE objective function. Under such circumstances, most of the optimization algorithms do not seem to be suited to face such problems as they usually require many evaluations before delivering an acceptable result.

As an alternative to traditional thresholding techniques, the problem of MT has also been handled through evolutionary methods. In general, they have demonstrated, under several circumstances, to deliver better results than those based on deterministic approaches in terms of accuracy and robustness [17, 18]. Under such conditions, recently, an extensive amount of evolutionary optimization approaches have been reported in the literature to find the appropriate threshold values by maximizing the complex objective function produced by Tsallis entropy. Such approaches have produced several interesting segmentation algorithms using different optimization methods such as Differential evolution (DE) [5], Particle Swarm Optimization algorithm (PSO) [19], Artificial Bee Colony (ABC) [16], Cuckoo Search algorithm (CSA) [13], Bacterial Foraging Optimization (BFOA) [12] and Harmony Search (HS) [20]. All these approaches permit with different results to optimize the TE fitness function in despite of its high multimodality characteristics.

However, one particular difficulty in their performance is the demand for a large number of fitness evaluations before delivering a satisfying result.

This chapter presents a multilevel thresholding method that uses the Electromagnetism-Like Algorithm (EMO) to find the best threshold values. EMO is a population-based evolutionary method which was firstly introduced by Birbil and Fang [21] to solve unconstrained optimization problems. The algorithm emulates the attraction–repulsion mechanism between charged particles within an electro-magnetism field. Each particle represents a solution and carries a certain amount of charge which is proportional to its fitness value. In turn, solutions are defined by position vectors which give real positions for particles within a multi-dimensional space. Moreover, objective function values of particles are calculated considering such position vectors. Each particle exerts repulsion or attraction forces over other members in the population; the resultant force acting over a particle is used to update its position. Clearly, the idea behind the EMO methodology is to move particles towards the optimum solution by exerting attraction or repulsion forces among them. Different to other evolutionary methods, EMO exhibits interesting search capabilities such as fast convergence still keeping its ability to avoid local minima in high modality environments [27–22]. Recent studies [30–23] demonstrate that the EMO algorithm presents the best balance between optimization results and demand of function evaluations. Such characteristics have attracted the attention of the evolutionary computation community, so that it has been effectively applied to solve a wide range of engineering problems such as flow-shop scheduling [24], communications [25], vehicle routing [26], array pattern optimization in circuits [27], neural network training [28], image processing [29] and control systems [30].

In this chapter, a new algorithm for multilevel segmentation based on the Electromagnetism-Like algorithm (EMO) is presented. In the approach, the EMO algorithm is used to find the optimal threshold values by maximizing the Tsallis entropy. As a result, the proposed algorithm can substantially reduce the number of function evaluations preserving the good search capabilities of an evolutionary method. In our approach, the algorithm uses as particles the encoding of a set of candidate threshold points. The TE objective function evaluates the segmentation quality of the candidate threshold points. Guided by the values of this objective function, the set of encoded candidate solutions are modified by using the EMO operators so that they can improve their segmentation quality as the optimization process evolves. In comparison to other similar algorithms, the proposed method deploys better segmentation results yet consuming less TE function evaluations.

The rest of the chapter is organized as follows. In Sect. 2.2, the standard EMO algorithm is introduced. Section 2.3 gives a simple description of the Tsallis entropy method. Section 2.4 explains the implementation of the proposed algorithm. Section 2.5 discusses experimental results and comparisons after testing the proposal over a set of benchmark images. Finally, in Sect. 2.6 the conclusions are discussed.

2.2   Electromagnetism—Like Optimization Algorithm (EMO)

EMO is a population-based evolutionary method which was firstly introduced by Birbil and Fang [31] to solve unconstrained optimization problems. Different to other evolutionary methods, EMO exhibits interesting search capabilities such as fast convergence still keeping its ability to avoid local minima in high modality environments [27–22]. Recent studies [30–23] demonstrate that the EMO algorithm presents the best balance between optimization results and demand of function evaluations. From the implementation point of view, EMO utilizes N different n-dimensional points $x_{i,t}$, $i = 1, 2, \ldots, n$, as a population for searching the feasible set $\mathbf{X} = \{x \in \mathbb{R}^n | l_i \leq x \leq u_i\}$, where t denotes the number of iteration (or generation) of the algorithm. The initial population $\mathbf{Sp}_t = \{x_{1,t}, x_{2,t}, \ldots, x_{N,t}\}$ (being $t = 1$), is taken of uniformly distributed samples of the search region, $\mathbf{X}$. We denote the population set at the t-th iteration by $\mathbf{Sp}_t$, and the members of $\mathbf{Sp}_t$ changes with t. After the initialization of $\mathbf{Sp}_t$, EMO continues its iterative process until a stopping condition (e.g. the maximum number of iterations) is met. An iteration of EMO consists of two main steps: in the first step, each point in $\mathbf{Sp}_t$ moves to a different location by using the attraction-repulsion mechanism of the electromagnetism theory [29]. In the second step, points moved by the electromagnetism principle are further perturbed locally by a local search and then become members of $\mathbf{Sp}_{t+1}$ in the $(t + 1)$-th iteration. Both the attraction-repulsion mechanism and the local search in EMO are responsible for driving the members, $x_{i,t}$, of $\mathbf{Sp}_t$ to the close proximity of the global optimum.

As with the electromagnetism theory for charged particles, each point $x_{i,t} \in \mathbf{Sp}_t$ in the search space $\mathbf{X}$ is assumed as a charged particle where the charge of a point is computed based on its objective function value. Points with better objective function value have higher charges than other points. The attraction-repulsion mechanism is a process in EMO by which points with more charge attract other points from $\mathbf{Sp}_t$, and points with less charge repel other points. Finally, a total force vector F_i^t, exerted on a point (e.g. the i-th point $x_{i,t}$) is calculated by adding these attraction—repulsion forces, and each $x_{i,t} \in \mathbf{Sp}_t$ is moved in the direction of its total force to the location $y_{i,t}$. A local search is used to explore the vicinity of the each particle according to its fitness. The members, $x_{i,t+1} \in \mathbf{Sp}_{t+1}$, of the $(t + 1)$-th iteration are then found by using:

$$x_{i,t+1} = \begin{cases} y_{i,t} & \text{if} \quad f(y_{i,t}) \leq f(z_{i,t}) \\ z_{i,t} & \text{otherwise} \end{cases} \tag{2.1}$$

Algorithm 2.1 shows the general scheme of EMO. We also provided the description of each step following the algorithm.

Algorithm 2.1 [EMO $(N, Iter_{max}, Iter_{local}, \delta)$]

1. Input parameters: Maximum number of iteration $Iter_{max}$, values for the local search parameter such $Iter_{local}$ and δ, and the size N of the population.
2. Initialize: set the iteration counter $t = 1$, initialize the number of $\mathbf{Sp}_t$ uniformly in $\mathbf{X}$ and identify the best point in $\mathbf{Sp}_t$.
3. while $t < Iter_{max}$ do
4. $F_i^t \leftarrow \text{CalcF}(\mathbf{Sp}_t)$
5. $y_{i,t} \leftarrow \text{Move}(x_{i,t}, F_i^t)$
6. $z_{i,t} \leftarrow \text{Local}(Iter_{local}, \delta, y_{i,t})$
7. $x_{i,t+1} \leftarrow \text{Select}(\mathbf{Sp}_{t+1}, y_{i,t}, z_{i,t})$
8. end while

Input parameters (Line 1): EMO algorithm runs for $Iter_{max}$ iterations. In the local search phase, $n \times Iter_{local}$ is the maximum number of locations $z_{i,t}$, within a δ distance of $y_{i,t}$, for each i dimension.

Initialize (Line 2): The points $x_{i,t}$, $t = 1$, are selected uniformly in $\mathbf{X}$, i.e. $x_{i,1} \sim Unif(\mathbf{X})$, $i = 1, 2, \ldots, N$, where $Unif$ represents the uniform distribution. The objective function values $f(x_{i,t})$ are computed, and the best point is identified for minimization:

$$x_t^B = \arg \min_{x_{i,t} \in \mathbf{S}_t} \{f(x_{i,t})\} \tag{2.2}$$

and for maximization:

$$x_t^B = \arg \max_{x_{i,t} \in \mathbf{S}_t} \{f(x_{i,t})\} \tag{2.3}$$

Calculate force (Line 4): In this step, a charged-like value $(q_{i,t})$ is assigned to each point $(x_{i,t})$. The charge $q_{i,t}$ of $x_{i,t}$ depends on $f(x_{i,t})$ and points with better objective function have more charge than others. The charges are computed as follows:

$$q_{i,t} = \exp\left(-n \frac{f(x_{i,t}) - f(x_t^B)}{\sum_{j=1}^N f(x_{i,t}) - f(x_t^B)}\right) \tag{2.4}$$

Then the force, $F_{i,j}^t$, between two points $x_{i,t}$ and $x_{j,t}$ is calculated using:

$$F_{i,j}^t = \begin{cases} (x_{j,t} - x_{i,t}) \frac{q_{i,t} \cdot q_{j,t}}{\|x_{j,t} - x_{i,t}\|^2} & \text{if } f(x_{i,t}) > f(x_{j,t}) \\ (x_{i,t} - x_{j,t}) \frac{q_{i,t} \cdot q_{j,t}}{\|x_{j,t} - x_{i,t}\|^2} & \text{if } f(x_{i,t}) \le f(x_{j,t}) \end{cases} \tag{2.5}$$

The total force, F_i^t, corresponding to $x_{i,t}$ is now calculated as:

$$F_i^t = \sum_{j=1, j \neq i}^{N} F_{i,j}^t \tag{2.6}$$

Move the point $x_{i,t}$ along F_i^t (Line 5): In this step, each point $x_{i,t}$ except for x_t^B is moved along the total force F_i^t using:

$$x_{i,t} = x_{i,t} + \lambda \frac{F_i^t}{\|F_i^t\|} (RNG), \quad i = 1, 2, \ldots, N; \; i \neq B \tag{2.7}$$

where $\lambda \sim Unif(0,1)$ for each coordinate of $x_{i,t}$, and RNG denotes the allowed range of movement toward the lower or upper bound for the corresponding dimension.

Local search (Line 6): For each $y_{i,t}$ a maximum of $iter_{local}$ points are generated in each coordinate direction in the δ neighbourhood of $y_{i,t}$. This means that the process of generating local point is continued for each $y_{i,t}$ until either a better $z_{i,t}$ is found or the $n \times Iter_{local}$ trial is reached.

Selection for the next iteration (Line 7): In this step, $x_{i,t+1} \in \mathbf{Sp}_{t+1}$ are selected from $y_{i,t}$ and $z_{i,t}$ using Eq. (2.1), and the best point is identified using Eq. (2.2) for minimization or Eq. (2.3) for maximization.

As it can be seen from Eqs. 2.1–2.8, the process to compute the elements of the new population $\mathbf{Sp}_{t+1}$ involves several operations that consider local and global information. Such process is more laborious than most of the evolutionary approaches which use only one equation to modify the individual position. This fact could be considered as an implementation disadvantage of the EMO method.

2.3 Tsallis Entropy (TE)

The entropy is defined in thermodynamic to measure the order of irreversibility in the universe. The concept of entropy physically expresses the amount of disorder of a system [13, 16]. In information theory, Shannon redefines the theory proposed by Boltzmann-Gibbs, and employ the entropy to measure the uncertainty regarding information of a system [16]. In other words, is possible the quantitatively measurement of the amount of information produced by a process.

The entropy in a discrete system takes a probability distribution defined as $p = \{p_i\}$, which represents the probability of find the system in a possible state i. Notice that $0 \leq p_i \leq 1$, $\sum_{i=1}^{k} p_i = 1$, and k is the total number of states. In addition, a physical or information system can be decomposed in two statistical independent subsystems A and B with probabilities p^A and p^B, where the probability of the composed system is given by $p^{A+B} = p^A \cdot p^B$. Such definition has been verified using the extensive property (additive) Eq. (2.8) proposed by Shannon [13, 16]:

$$S(A+B) = S(A) + S(B) \qquad (2.8)$$

Tsallis proposed a generalized form of statistics based on the related concepts and the multi-fractal theory. The Tsallis entropic form is an important tool used to describe the thermo statistical properties of non-extensive systems and is defined as:

$$S_q = \frac{1 - \sum_{i=1}^{k} (p_i)^q}{q - 1} \qquad (2.9)$$

where S is the Tsallis entropy, q is the Tsallis entropic index that represents the degree of non-extensivity and k is the total number of possibilities of the system. Since Tsallis entropy is non-extensive, it is necessary to redefine the additive entropic rule of Eq. (2.8).

$$S_q(A+B) = S_q(A) + S_q(B) + (1-q) \cdot S_q(A) \cdot S_q(B) \qquad (2.10)$$

Since image segmentation has non-additive information content, it is possible to use the Tsallis entropy to find the best thresholds [16]. A digital gray scale image has k gray levels that are defined by the histogram. The easiest thresholding considers to classes divided by a one threshold (bi-level), to solve this problem is considered the probability distribution of the gray levels ($p_i = p_1, p_2, \ldots p_k$). For each class A and B two probability distributions are created Eq. (2.11)

$$p_A = \frac{p_1}{P^A}, \frac{p_2}{P^A}, \ldots \frac{p_{th}}{P^A} \quad \text{and} \quad p_B = \frac{p_1}{P^B}, \frac{p_2}{P^B}, \ldots \frac{p_k}{P^B} \qquad (2.11)$$

where

$$P^A = \sum_{i=1}^{th} p_i \quad \text{and} \quad P^B = \sum_{i=th+1}^{k} p_i \qquad (2.12)$$

The TE for class A and Class B is defined as follows:

$$S_q^A(th) = \frac{1 - \sum_{i=1}^{th} \left(\frac{p_i}{P^A}\right)^q}{q-1}, \quad S_q^B(th) = \frac{1 - \sum_{i=th+1}^{k} \left(\frac{p_i}{P^B}\right)^q}{q-1} \qquad (2.13)$$

TE value depends directly on the selected threshold value, and it maximizes the information measured between two classes. If the value of $S_q(th)$ is maximized it means that th is the optimal value. In order to verify the efficiency of the selected th, in Eq. (2.14) is proposed an objective function using Eq. (2.10):

$$TH_{opt}(th) = \arg\max\left(S_q^A(th) + S_q^B(th) + (1-q) \cdot S_q^A(th) \cdot S_q^B(th)\right) \qquad (2.14)$$

The previous description of this bi-level method can be extended for the identification of multiple thresholds. Considering nt thresholds, it is possible separate the original image into $(nt\text{-}1)$ classes. Under the segmentation approach, the optimization problem turns into a multidimensional situation. The candidate solutions are conformed as $\mathbf{th}^j = [th_1, th_2, \ldots th_{nt}]$. For each class is computed the entropy using the Tsallis methodology and the objective function is redefined as follows:

$$TH_{opt}(\mathbf{th}) = \arg\max(L)$$
$$L = S_q^1(th_1) + S_q^2(th_2) + \cdots + S_q^{nt}(th_{nt}) + (1-q) \cdot S_q^1(th_1) \cdot S_q^2(th_2)\ldots S_q^{nt}(th_{nt})$$

$$(2.15)$$

where

$$S_q^1(th_1) = \frac{1 - \sum_{i=1}^{th_1} \left(\frac{p_i}{P^1}\right)^q}{q-1}, \quad S_q^2(th) = \frac{1 - \sum_{i=th_1+1}^{th_2} \left(\frac{p_i}{P^2}\right)^q}{q-1}, \ldots,$$
$$S_q^{nt}(th) = \frac{1 - \sum_{i=th_{nt-1}+1}^{th_{nt}} \left(\frac{p_i}{P^{nt}}\right)^q}{q-1}$$

$$(2.16)$$

Notice that for each threshold the entropy is computed and corresponds to a specific class. However there exist an extra class it means that exist $nt+1$ classes. The extra class is considered *default* class because it is computed from nt to k Eq. (2.17).

$$S_q^{def}(th_k) = \frac{1 - \sum_{i=th_{nt}+1}^{k} \left(\frac{p_i}{P^k}\right)^q}{q-1}$$

$$(2.17)$$

From Eq. 2.14, it is evident that TE presents a simple functional formulation for bi-level thresholding. However, as it is shown by Eqs. 2.15 and 2.16, when it is considered multilevel thresholding (MT), its evaluation becomes computationally expensive, since each threshold point adds restrictions, multimodality and complexity to its functional formulation. Therefore, in the process of finding the appropriate threshold values, it is desired to limit the number of evaluations of the TE objective function. Under such circumstances, most of the optimization algorithms do not seem to be suited to face such problems as they usually require many evaluations before delivering an acceptable result.

2.4 Multilevel Thresholding Using EMO and Tsallis Entropy (TSEMO)

In this chapter, a new algorithm for multilevel segmentation based on the Electromagnetism-Like algorithm (EMO) is presented. In the approach, the EMO algorithm is used to find the optimal threshold values by maximizing the complex Tsallis entropy. Different to other evolutionary methods, EMO exhibits interesting search capabilities such as fast convergence still keeping its ability to avoid local minima in high modality environments [27–22]. Recent studies [30–23] demonstrate that the EMO algorithm presents the best balance between optimization results and demand of function evaluations. As a result, the proposed segmentation algorithm can substantially reduce the number of function evaluations preserving the good search capabilities of an evolutionary method. However, as it can be seen from Eqs. 2.1–2.8, the process of EMO, to compute the elements of the new population, involves several operations that consider local and global information. Such process is more laborious than most of the evolutionary approaches which use only one equation to modify the individual position. This fact could be considered as an implementation disadvantage of the EMO method. In this section, the proposed approach is discussed

2.4.1 Particle Representation

Each particle uses nt decision variables in the optimization algorithm. Such elements represent a different threshold point used for the segmentation. Therefore, the complete population is represented as:

$$\mathbf{Sp}_t = [\mathbf{th}_1, \mathbf{th}_2, \ldots, \mathbf{th}_N], \quad \mathbf{th}_i = [th_1, th_2, \ldots, th_{nt}]^T \qquad (2.18)$$

where t represents the iteration number, T refers to the transpose operator, N is the size of the population.

2.4.2 EMO Implementation

The proposed segmentation algorithm has been implemented considered the Tsallis pseudo-additive entropic rule as objective function Eq. (2.15). The implementation of the EMO algorithm can be summarized into the following steps:

Step 1   Read the image I and store it into I_{Gr}.
Step 2   Obtain histogram h^{Gr} of I_{Gr}.
Step 3   Initialize the EMO parameters: $Iter_{max}$, $Iter_{local}$, δ, k and N.
Step 4   Initialize a population $\mathbf{Sp}_t$ of N random particles with nt dimensions.
Step 5   Compute the Tsallis entropy $S_q^i(\mathbf{Sp}_t)$ for each element of $\mathbf{Sp}_t$, Eqs. (2.16) and (2.17). Evaluate $\mathbf{Sp}_t$ in the objective function $TH_{opt}(\mathbf{Sp}_t)$ Eq. (2.15).
Step 6   Compute the charge of each particle using Eq. (2.4), and with Eqs. (2.5) and (2.6) compute the total force vector.
Step 7   Move the entire population $\mathbf{Sp}_t$ along the total force vector using Eq. (2.7).
Step 8   Apply the local search to the moved population and select the best elements of this search based on their objective function values.
Step 9   The t index is increased in 1, If $t \geq Iter_{max}$ or if the stop criteria is satisfied the algorithm finishes the iteration process and jump to step 11. Otherwise jump to step 7.
Step 10 Select the particle that has the best $x_t^{B^c}$ objective function value Eqs. (2.3) and (2.15).
Step 11 Apply the thresholds values contained in $x_t^{B^c}$ to the image I_{Gr}.

2.4.3 Multilevel Thresholding

Ones the EMO algorithm finds the best threshold values that maximize the objective function. These are used to segment the image pixels. There exist several ways to apply the thresholds, in this chapter we use the following rule for two levels:

$$I_s(r,c) = \begin{cases} I_{Gr}(r,c) & \text{if} & I_{Gr}(r,c) \leq th_1 \\ th_1 & \text{if} & th_1 < I_{Gr}(r,c) \leq th_2 \\ I_{Gr}(r,c) & \text{if} & I_{Gr}(r,c) > th_2 \end{cases} \qquad (2.19)$$

where $I_s(r,c)$ is the gray value of the segmented image, $I_{Gr}(r,c)$ is the gray value of the original image both in the pixel position r,c. th_1 and th_2 are the threshold values obtained by the EMO approach. Equation (2.19) can be easily extended for more than two levels Eq. (2.20).

$$I_s(r,c) = \begin{cases} I_{Gr}(r,c) & \text{if} \quad I_{Gr}(r,c) \leq th_1 \\ th_{i-1} & \text{if} \quad th_{i-1} < I_{Gr}(r,c) \leq th_i, \quad i = 2,3,\ldots nt-1 \\ I_{Gr}(r,c) & \text{if} \quad I_{Gr}(r,c) > th_{nt} \end{cases} \qquad (2.20)$$

2.5 Experimental Results

The proposed algorithm has been tested under a set of 11 benchmark images. Some of these images are widely used in the image processing literature to test different methods (Lena, Cameraman, Hunter, Baboon, etc.) [13]. All the images have the same size (512×512 pixels) and they are in JPGE format.

In order to carry out the algorithm analysis the proposed TSEMO is compared to state-of-the-art thresholding methods, such Cuckoo Search algorithm (CSA) [13] and Particle Swarm Optimization (PSO) [17]. Since all the methods are stochastic, it is necessary to employ statistical metrics to compare the efficiency of the algorithms. Hence, all algorithms are executed 35 times per image, according to the related literature the number the thresholds for test are $th = 2, 3, 4, 5$ [13, 19]. In each experiment the stop criteria is set to 50 iterations. In order to verify the stability at the end of each test the standard deviation (STD) is obtained Eq. (2.21). If the STD value increases the algorithms becomes more instable [29].

$$STD = \sqrt{\sum_{i=1}^{Iter_{\max}} \frac{(\sigma_i - \mu)}{Ru}} \tag{2.21}$$

On the other hand the peak-to-signal ratio ($PSNR$) is used to compare the similarity of an image (image segmented) against a reference image (original image) based on the mean square error (MSE) of each pixel [5, 13, 30, 32]. Both $PSNR$ and MSE are defined as:

$$PSNR = 20 \log_{10}\left(\frac{255}{RMSE}\right), \quad \text{(dB)}$$

$$RMSE = \sqrt{\frac{\sum_{i=1}^{ro} \sum_{j=1}^{co} \left(I_{Gr}(i,j) - I_{th}(i,j)\right)}{ro \times co}} \tag{2.22}$$

where I_{Gr} is the original image, I_{th} is the segmented image and ro, co are the total number of rows and columns of the image, respectively. The Structure Similarity Index ($SSIM$) is used to compare the structures of the original umbralized image [33] it is defined in Eq. (2.23). A higher $SSIM$ value means that the performance of the segmentation is better.

$$SSIM(I_{Gr}, I_{th}) = \frac{\left(2\mu_{I_{Gr}}\mu_{I_{th}} + C1\right)\left(2\sigma_{I_{Gr}I_{th}} + C2\right)}{\left(\mu_{I_{Gr}}^2 + \mu_{I_{th}}^2 + C1\right)\left(\sigma_{I_{Gr}}^2 + \sigma_{I_{th}}^2 + C2\right)}$$

$$\sigma_{I_oI_{Gr}} = \frac{1}{N-1}\sum_{i=1}^{N}\left(I_{Gr_i} + \mu_{I_{Gr}}\right)\left(I_{th_i} + \mu_{I_{th}}\right) \tag{2.23}$$

where $\mu_{I_{Gr}}$ and $\mu_{I_{th}}$ are the mean value of the original and the umbralized image respectively, for each image the values of $\sigma_{I_{Gr}}$ and $\sigma_{I_{th}}$ corresponds to the standard deviation. $C1$ and $C2$ are constants used to avoid the instability when

$\mu_{I_{Gr}}^2 + \mu_{I_{th}}^2 \approx 0$, experimentally in [12] both values are $C1 = C2 = 0.065$. Another method used to measure the quality of the segmented image is the Feature Similarity Index (*FSIM*) [34]. *FSIM* calculates the similarity between two images, in this cases the original gray scale image and the segmented image Eq. (2.24). As *PSNR* and *SSIM* the higher value is interpreted as better performance of the thresholding method.

$$FSIM = \frac{\sum_{w \in \Omega} S_L(w) PC_m(w)}{\sum_{w \in \Omega} PC_m(w)} \tag{2.24}$$

where Ω represents the entire domain of the image:

$$
\begin{aligned}
S_L(w) &= S_{PC}(w) S_G(W) \\
S_{PC}(w) &= \frac{2PC_1(w)PC_2(w) + T_1}{PC_1^2(w) + PC_2^2(w) + T_1} \\
S_G(W) &= \frac{2G_1(w)G_2(w) + T_2}{G_1^2(w) + G_2^2(w) + T_2}
\end{aligned}
\tag{2.25}
$$

G is the gradient magnitude (GM) of an image and is defined as:

$$G = \sqrt{G_x^2 + G_y^2} \tag{2.26}$$

PC is the phase congruence:

$$PC(w) = \frac{E(w)}{\left(\varepsilon + \sum_n A_n(w)\right)} \tag{2.27}$$

$A_n(w)$ is the local amplitude on scale n and $E(w)$ is the magnitude of the response vector in w on n. ε is an small positive number and $PC_m(w) = \max(PC_1(w), PC_2(w))$. On the other hand, Table 2.1 presents the parameters for the EMO algorithm. They have been obtained using the criterion proposed in [31] and kept for all test images.

2.5.1 Tsallis Entropy Results

In this section, the results of the TSEMO algorithm are analyzed, considering as objective function Eq. (2.15) the Tsallis entropy [35]. The approach is applied over

Table 2.1 EMO parameters

$Iter_{max}$	$Iter_{local}$	δ	N
200	10	0.25	50

the complete set of benchmark images whereas the results are registered in Table 2.2. Such results present the best threshold values obtained after testing the TSEMO algorithm, considering four different threshold points $th = 2, 3, 4, 5$. In Table 2.2, it is also shown the *PSNR*, *STD*, *SSIM* and *FSIM* values.

There have been selected five images of the set to show (graphically) the segmentation results. Figure 2.1 presents the images selected from the benchmark set and their respective histograms which possess irregular distributions (particularly Fig. 2.1j). Under such circumstances, classical methods face great difficulties to find the best threshold values.

Table 2.3 shows the images obtained after processing 5 original images selected from the entire benchmark set, applying the proposed algorithm. The results present the segmented images considering four different threshold levels $th = 2, 3, 4, 5$. In Table 2.3, it is also shown the evolution of the objective function during one execution. From the results, it is possible to appreciate that the TSEMO converges (stabilizes) around the first 100 iterations. The segmented images provide evidence that the outcome is better with $th = 4$ and $th = 5$; however, if the segmentation task does not requires to be extremely accurate then it is possible to select $th = 3$.

2.5.2 Comparisons

In order to demonstrate that the TSEMO is an interesting alternative for MT, the proposed algorithm is compared with two state-of-the-art implementations. The methods used for comparison are: the Cuckoo Search Algorithm (CSA) [13] and the Particle Swarm Optimization (PSO) [19], both methods uses the Tsallis entropy.

All the algorithms run 35 times over each selected image. The images used for this test are the same of the selected in Sect. 2.5.1 (Camera man, Lena, Baboon, Hunter and Butterfly). For each image is computed the *PSNR*, *STD*, *SSIM*, *FSIM* values and the mean of the objective function.

The comparison results between the three methods are divided in two tables, Table 2.4 shows the *STD* and mean values of the fitness function. Table 2.5 presents the values of the quality metrics obtained after apply the thresholds over the test images.

The fitness values of four methods are statistically compared using a non-parametric significance proof known as the Wilcoxon's rank test [36] that is conducted with 35 independent samples. Such proof allows assessing result differences among two related methods. The analysis is performed considering a 5% significance level over the best fitness (Tsallis entropy) value data corresponding to the five threshold points. Table 2.6 reports the p-values produced by Wilcoxon's test for a pair-wise comparison of the fitness function between two groups formed as TSEMO versus CSA, TSEMO versus PSO. As a null hypothesis, it is assumed

Table 2.2 Result after applying the MTEMO to the set of benchmark images

Image	k	Thresholds x_t^B	PSNR	STD	SSIM	FSIM
Camera man	2	71, 130	23.1227	31.00 E–04	0.9174	0.8901
	3	71, 130, 193	18.0122	72.01 E–04	0.8875	0.8456
	4	44, 84, 120, 156	24.9589	86.01 E–03	0.9363	0.9149
	5	44, 84, 120, 156, 196	23.0283	7.90 E–01	0.9289	0.8960
Lena	2	79, 127	23.9756	7.21 E–05	0.9083	0.8961
	3	79, 127, 177	21.0043	14.37 E–04	0.8660	0.8197
	4	62, 94, 127, 161	24.0020	18.69 E–03	0.9057	0.8851
	5	62, 94, 127, 161, 194	23.3736	39.82 E–02	0.8956	0.8684
Baboon	2	15, 105	23.5906	18.51 E–06	0.9480	0.9437
	3	51, 105, 158	19.9394	28.78 E–02	0.9011	0.9059
	4	33, 70, 107, 143	23.5022	22.65 E–02	0.9530	0.9594
	5	33, 70, 107, 143, 179	21.9540	37.13 E–01	0.9401	0.9417
Hunter	2	60, 119	22.8774	17.89 E–04	0.9192	0.8916
	3	60, 119, 179	20.2426	54.12 E–04	0.9031	0.8652
	4	46, 90, 134, 178	22.4723	1.94 E–02	0.9347	0.9159
	5	46, 90, 134, 178, 219	22.4025	1.23 E–01	0.9349	0.9173
Airplane	2	69, 125	25.4874	17.31 E–04	0.9685	0.9239
	3	69, 125, 180	22.9974	17.89 E–04	0.9433	0.8909
	4	55, 88, 122, 155	28.5400	19.21 E–03	0.9848	0.9677
	5	55, 88, 122, 155, 188	26.4997	35.08 E–03	0.9663	0.9417
Peppers	2	70, 145	19.6654	54.83 E–02	0.8697	0.8378
	3	70, 145, 223	17.2736	1.31 E–01	0.8437	0.7534
	4	46, 88, 132, 175	21.8275	3.02 E–04	0.8976	0.8552
	5	46, 88, 132, 175, 223	21.1207	6.34 E–03	0.8976	0.8304
Living room	2	55, 111	22.6665	47.11 E–03	0.9116	0.8966
	3	55, 111, 179	18.0379	15.27 E–04	0.8482	0.8132
	4	42, 85, 124, 162	21.7235	93.35 E–03	0.9170	0.9090
	5	42, 85, 124, 162, 201	21.3118	94.32 E–03	0.9183	0.9029
Blonde	2	62, 110	25.8389	31.91 E–04	0.9645	0.9503
	3	62, 110, 155	21.5001	37.05 E–04	0.9012	0.8759
	4	36, 65, 100, 134	25.9787	17.45 E–03	0.9606	0.9491
	5	36, 65, 100, 134, 168	23.1835	48.20 E–03	0.9328	0.9077
Bridge	2	65, 131	20.1408	22.71 E–04	0.8619	0.8749
	3	65, 131, 191	18.7016	40.49 E–04	0.8410	0.8479
	4	45, 88, 131, 171	21.4247	38.48 E–03	0.9168	0.9279
	5	45, 88, 131, 171, 211	21.0157	66.16 E–03	0.9153	0.9217
Butterfly	2	83, 120	26.7319	96.11 E–03	0.9493	0.9195
	3	83, 120, 156	24.4582	39.04 E–03	0.9386	0.8934
	4	70, 94, 119, 144	27.0221	14.59 E–02	0.9653	0.9417
	5	70, 94, 119, 144, 172	25.7809	98.61 E–02	0.9610	0.9283
Lake	2	71, 121	27.8565	10.69 E–04	0.9729	0.9638
	3	71, 121, 173	23.7695	12.87 E–04	0.9399	0.9288
	4	41, 80, 119, 159	24.7454	11.97 E–03	0.9587	0.9422
	5	41, 80, 119, 159, 197	22.4347	11.80 E–03	0.9439	0.9213

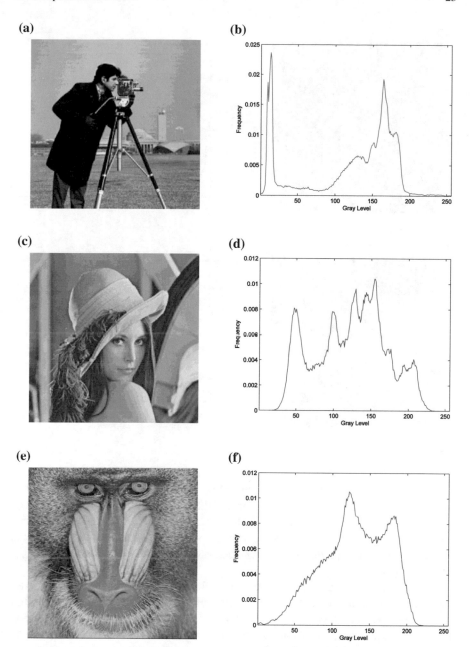

Fig. 2.1 **a** Camera man, **c** Lena, **e** Baboon, **g** Hunter and **i** Butterfly, the selected benchmak images. **b, d, f, h, j** histograms of the images

(g)

(h)

(i)

(j)

Fig. 2.1 (continued)

that there is no difference between the values of the two algorithms tested. The alternative hypothesis considers an existent difference between the values of both approaches. All p-values reported in Table 2.6 are less than 0.05 (5% significance level) which is a strong evidence against the null hypothesis, indicating that the TSEMO fitness values for the performance are statistically better and it has not occurred by chance.

On the other hand, to compare the fitness of the three methods Table 2.7 shows the fitness values obtained for the reduced set of image (5 images). Each algorithm runs 1000 times and the best value of each run is stored, at the end of the evolution process the best stored values are plotted. From Table 2.6 it is possible to analyze that TSEMO and CSA reach the maximum entropy values in less iterations than the PSO method.

Table 2.3 Results after appliying the MT-EMO using Tsallis entropy over the selected benchamark images

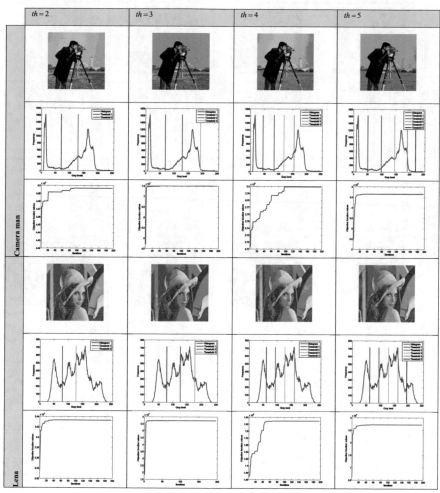

(continued)

Table 2.3 (continued)

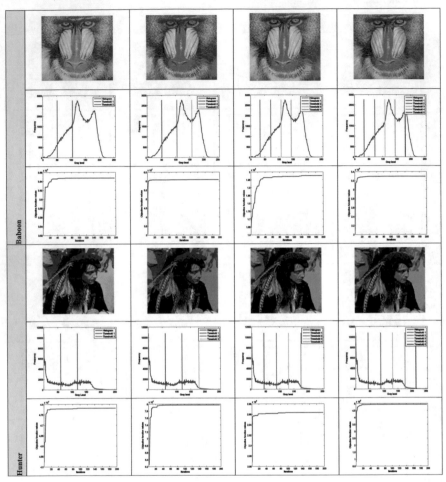

Table 2.3 (continued)

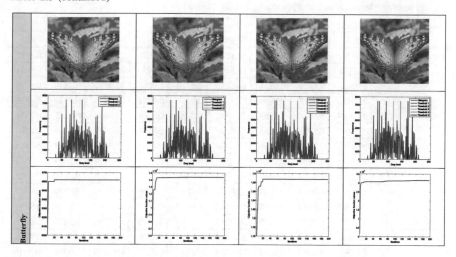

Table 2.4 Comparison of the *STD* and mean values of the TSEMO, CSA and PSO applied over the selected test images using Tsallis method

Image	k	TSEMO		CSA		PSO	
		STD	Mean	*STD*	Mean	*STD*	Mean
Camera man	2	31.00 E–04	4.49 E+04	89.56 E–04	4.02 E+04	83.00 E02	4.19 E+04
	3	72.01 E–04	7.49 E+04	98.32 E–04	6.99 E+04	89.00 E+00	7.27 E+04
	4	86.01 E–03	2.79 E+06	18.68 E–03	2.18 E+06	12.35 E+02	2.37 E+06
	5	7.90 E–01	4.65 E+06	69.98 E–01	4.56 E+06	5.38 E+03	4.28 E+06
Lena	2	7.21 E–05	3.43 E+04	2.61 E+00	3.33 E+04	15.27 E+00	3.30 E+04
	3	14.37 E–04	5.72 E+04	3.39 E+00	5.67 E+04	3.31 E+00	5.62 E+04
	4	18.69 E–03	1.62 E+06	5.52 E+00	1.45 E+06	7.35 E+00	1.45 E+06
	5	39.82 E–02	2.71 E+06	8.50 E+01	2.55 E+06	2.92 E+00	2.59 E+06
Baboon	2	18.51 E–06	3.64 E+04	15.11 E–02	3.47 E+04	2.64 E+00	3.40 E+04
	3	28.78 E–02	6.08 E+04	40.80 E–02	6.05 E+04	1.44 E+00	6.03 E+04
	4	22.65 E–02	1.97 E+06	62.02 E–02	1.90 E+06	8.11 E+00	1.86 E+06
	5	37.13 E–01	3.29 E+06	52.74 E–02	3.20 E+06	2.68 E+00	3.20 E+06
Hunter	2	17.89 E–04	4.78 E+04	7.38 E–04	4.70 E+04	4.38 E+00	4.72 E+04
	3	54.12 E–04	7.97 E+04	2.95 E–04	7.89 E+04	9.47 E+00	7.85 E+04
	4	1.94 E–02	2.96 E+06	1.62 E–01	2.93 E+06	1.04 E+01	2.92 E+04
	5	1.23 E–01	4.94 E+06	2.46 E–01	4.89 E+06	3.23 E+02	4.75 E+04
Butterfly	2	96.11 E–03	8.61 E+03	12.78 E–02	8.56 E+03	6.36 E–01	8.55 E+03
	3	39.04 E–03	1.43 E+04	19.00 E–02	1.38 E+04	11.56 E–01	1.35 E+04
	4	14.59 E–02	1.88 E+05	11.04 E–01	1.80 E+05	1.04 E+00	1.81 E+05
	5	98.61 E–02	3.14 E+05	1.58 E+00	3.07 E+05	3.58 E+00	2.96 E+05

Table 2.5 Comparison of the *PSNR*, *SSIM* and *FSIM* values of the TSEMO, CSA and PSO applied over the selected test images using Tsallis method

Image	k	TSEMO			CSA			PSO		
		PSNR	*SSIM*	*FSIM*	*PSNR*	*SSIM*	*FSIM*	*PSNR*	*SSIM*	*FSIM*
Camera man	2	23.1227	0.9174	0.8901	23.1194	0.9173	0.8901	22.9737	0.9160	0.8871
	3	18.0998	0.8875	0.8509	18.7480	0.8918	0.8456	18.0122	0.8874	0.8441
	4	25.0021	0.9369	0.9151	24.5479	0.9349	0.9097	23.3230	0.9280	0.8976
	5	22.9136	0.9286	0.8950	22.5284	0.9243	0.8891	21.9598	0.9222	0.8839
Lena	2	23.9982	0.9088	0.8966	23.9756	0.9083	0.8961	23.9594	0.9085	0.8953
	3	21.2592	0.8699	0.8255	20.9669	0.8655	0.8192	20.9989	0.8659	0.8196
	4	23.9783	0.9056	0.8849	23.9493	0.9056	0.8846	23.8175	0.9032	0.8815
	5	23.4275	0.8954	0.8691	23.3099	0.8960	0.8689	23.3777	0.8949	0.8674
Baboon	2	23.7510	0.9496	0.9452	23.5906	0.9480	0.9410	23.5048	0.9475	0.9323
	3	19.9386	0.9007	0.9057	19.9031	0.8810	0.8759	19.8021	0.8729	0.8729
	4	23.5165	0.9532	0.9593	23.5106	0.9270	0.9295	23.5163	0.9125	0.9159
	5	22.0538	0.9410	0.9408	21.9071	0.9399	0.9112	21.7165	0.9350	0.9377
Hunter	2	22.8783	0.9192	0.8916	22.8074	0.9089	0.8826	22.7910	0.9093	0.8818
	3	20.2581	0.9034	0.8654	20.0026	0.8931	0.8552	20.0858	0.8921	0.8521
	4	22.4221	0.9341	0.9159	21.3972	0.9237	0.9055	21.5061	0.9244	0.9024
	5	22.5014	0.9355	0.9199	21.3171	0.9236	0.9063	21.3754	0.9254	0.9005
Butterfly	2	26.8352	0.9504	0.9212	25.7319	0.9493	0.9195	25.1635	0.9431	0.9150
	3	24.4144	0.9383	0.8926	23.4545	0.9300	0.8834	23.5251	0.9315	0.8846
	4	27.1226	0.9653	0.9420	26.0314	0.9653	0.9317	26.0810	0.9653	0.9321
	5	25.8838	0.9609	0.9285	24.0086	0.9516	0.9201	24.4870	0.9533	0.9142

Table 2.6 Wilcoxon *p*-values of the compared algorithm TSEMO versus CSA and TSEMO verssus PSO

Image	k	*p*-values	
		TSEMO versus CS	TSEMO versus PSO
Camera man	2	6.2137 E–07	8.3280 E–06
	3	1.0162 E–07	2.0000 E–03
	4	8.8834 E–08	13.710 E–03
	5	16.600 E–03	50.600 E–03
Lena	2	3.7419 E–08	1.6604 E–04
	3	1.4606 E–06	1.3600 E–02
	4	1.2832 E–07	2.9000 E–03
	5	3.9866 E–05	8.9000 E–03
Baboon	2	1.5047 E–06	2.5500 E–02
	3	6.2792 E–05	5.1000 E–03
	4	2.1444 E–12	3.3134 E–05
	5	2.1693 E–11	1.8000 E–03

(continued)

Table 2.6 (continued)

Image	k	p-values	
		TSEMO versus CS	TSEMO versus PSO
Hunter	2	2.2100 E–02	2.2740 E–02
	3	3.6961 E–04	1.1500 E–02
	4	6.8180 E–02	9.9410 E–09
	5	5.8200 E–02	2.4939 E–04
Airplane	2	3.0000 E–03	6.6300 E–03
	3	7.6000 E–03	3.5940 E–02
	4	4.8092 E–12	1.1446 E–06
	5	1.0023 E–09	2.7440 E–02
Peppers	2	2.7419 E–04	1.3194 E–04
	3	2.6975 E–08	3.5380 E–02
	4	1.5260 E–08	6.0360 E–02
	5	7.2818 E–08	7.6730 E–02
Living room	2	1.4000 E–03	2.6340 E–02
	3	6.8066 E–08	2.8000 E–03
	4	8.7456 E–07	5.8730 E–03
	5	1.7000 E–03	5.1580 E–03
Blonde	2	3.0000 E–03	4.1320 E–02
	3	5.9000 E–03	8.9300 E–02
	4	1.3800 E–02	2.7700 E–02
	5	2.3440 E–02	5.6000 E–03
Bridge	2	1.5000 E–03	1.5700 E–02
	3	1.4300 E–02	1.5350 E–02
	4	1.7871 E–06	7.0400 E–03
	5	8.7000 E–03	1.2400 E–02
Butterfly	2	1.5000 E–03	1.1150 E–02
	3	3.1800 E–02	1.3760 E–02
	4	4.8445 E–07	8.1800 E–03
	5	1.6000 E–02	1.0630 E–02
Lake	2	7.6118 E–06	2.9500 E–02
	3	1.2514 E–06	6.5644 E–06
	4	2.2366 E–10	6.6000 E–03
	5	5.3980 E–06	9.4790 E–03

Table 2.7 Fitness comparsion of PSO (blue line), CSA (Black line) and EMO (red line) applied for multilevel tresholding using TE

2.6   Conclusions

In this chapter, a new algorithm for multilevel segmentation based on the Electromagnetism-Like algorithm (EMO) has been presented. The proposed approach considers the segmentation process as an optimization problem, where EMO is employed to find the optimal threshold points that maximize the Tsallis entropy (TE). As a result, the proposed algorithm can substantially reduce the number of function evaluations preserving the good search capabilities of an evolutionary method. In our approach, the algorithm uses as particles the encoding of a set of candidate threshold points. The TE objective function evaluates the segmentation quality of the candidate threshold points. Guided by the values of this objective function, the set of encoded candidate solutions are modified by using the EMO operators so that they can improve their segmentation quality as the optimization process evolves. In order to evaluate the quality of the segmented images, the use of the *PSNR*, *STD*, *SSIM* and *FSIM* is proposed. Such metrics considers the coincidences between the original and the segmented image.

The study compares the proposed approach with other two similar approaches the Cuckoo Search algorithm (CSA) and Particle Swarm Optimization algorithm (PSO). The efficiency of the algorithms is evaluated in terms of *PSNR*, *STD*, *SSIM*, *FSIM* and fitness values. Such comparisons provide evidence of the accuracy, convergence and robustness of the proposed approach. The fitness of TSEMO is compared with the CSA and PSO where is possible to see that both EMO and CSA need a reduced number of iterations to converge. However the speed of convergence of EMO is higher than de CSA in the same way PSO is the slower and it has lack of accuracy. Although the results offer evidence to demonstrate that the TSEMO method can yield good results on complicated images, the aim of our chapter is not to devise a multilevel thresholding algorithm that could beat all currently available methods, but to show that electro-magnetism systems can be effectively considered as an attractive alternative for this purpose.

References

1. Cuevas, E., Zaldivar, D., Pérez-Cisneros, M., Seeking multi-thresholds for image segmentation with Learning Automata, Machine Vision and Applications, 22 (5), (2011), 805–818.
2. Y. Kong, Y. Deng, Q. Dai, and S. Member, "Discriminative Clustering and Feature Selection for Brain MRI Segmentation," IEEE Signal Process. Lett., vol. 22, no. 5, pp. 573–577, 2015.
3. X. Cao, Q. Li, X. Du, M. Zhang, and X. Zheng, "Exploring effect of segmentation scale on orient-based crop identification using HJ CCD data in Northeast China," *IOP Conf. Ser. Earth Environ. Sci.*, vol. 17, p. 012047, 2014.
4. A. K. Bhandari, V. K. Singh, A. Kumar, and G. K. Singh, "Cuckoo search algorithm and wind driven optimization based study of satellite image segmentation for multilevel thresholding using Kapur's entropy," *Expert Syst. Appl.*, vol. 41, no. 7, pp. 3538–3560, 2014.
5. S. Sarkar and S. Das, "Multilevel Image Thresholding Based on 2D Histogram and Maximum Tsallis Entropy—A Differential Evolution Approach," *Lect. Notes Comput. Sci.*, vol. 22, no. 12, pp. 4788–4797, 2013.
6. B. Akay, "A study on particle swarm optimization and artificial bee colony algorithms for multilevel thresholding," *Appl. Soft Comput.*, vol. 13, no. 6, pp. 3066–3091, 2012.
7. H. Xia, S. Song, and L. He, "A modified Gaussian mixture background model via spatiotemporal distribution with shadow detection," *Signal, Image Video Process.*, 2015.
8. G. Moser, S. B. Serpico, and S. Member, "Generalized Minimum-Error Thresholding for Unsupervised Change Detection From SAR Amplitude Imagery.pdf," *IEEE Trans. Geosci. Remote Sens.*, vol. 44, no. 10, pp. 2972–2982, 2006.
9. Sezgin M, "Survey over image thresholding techniques and quantitative performance evaluation," *J. Electron. Imaging*, vol. 13, no. January, pp. 146–168, 2004.
10. N. Otsu, "A Threshold Selection Method from Gray-Level Histograms," *IEEE Trans. Syst. Man. Cybern.*, vol. 9, no. 1, pp. 62–66, 1979.
11. A. K. C. J. N. Kapur, P. K. Sahoo, A. K. C. Wong, "A new method for gray-level picture thresholding using the entropy of the histogram." Computer Vision Graphics Image Processing, pp. 273–285, 1985.
12. P. D. Sathya and R. Kayalvizhi, "Optimal multilevel thresholding using bacterial foraging algorithm," *Expert Syst. Appl.*, vol. 38, no. 12, pp. 15549–15564, 2011.

13. S. Agrawal, R. Panda, S. Bhuyan, and B. K. Panigrahi, "Tsallis entropy based optimal multilevel thresholding using cuckoo search algorithm," *Swarm Evol. Comput.*, vol. 11, pp. 16–30, 2013.
14. C. Tsallis, "Possible generalization of Boltzmann-Gibbs statistics," *J. Stat. Phys.*, vol. 52, pp. 479–487, 1988.
15. E. K. Tang, P. N. Suganthan, and X. Yao, "An analysis of diversity measures," *Mach. Learn.*, vol. 65, no. April, pp. 247–271, 2006.
16. Y. Zhang and L. Wu, "Optimal multi-level thresholding based on maximum Tsallis entropy via an artificial bee colony approach," *Entropy*, vol. 13, pp. 841–859, 2011.
17. C. Tsallis, "Computational applications of nonextensive statistical mechanics," *J. Comput. Appl. Math.*, vol. 227, no. 1, pp. 51–58, 2009.
18. Cuevas, E., Ortega-Sánchez, N., Zaldivar, D., Pérez-Cisneros, M., Circle detection by Harmony Search Optimization, Journal of Intelligent and Robotic Systems: Theory and Applications, 66(3), (2012), 359–376.
19. N. Sri, M. Raja, G. Kavitha, and S. Ramakrishnan, "Analysis of Vasculature in Human Retinal Images Using Particle Swarm Optimization Based Tsallis Multi-level Thresholding and Similarity Measures," *Lect. Notes Comput. Sci. (including Subser. Lect. Notes Artif. Intell. Lect. Notes Bioinformatics)*, vol. 7677, no. 1, pp. 380–387, 2012.
20. Oliva, D., Cuevas, E., Pajares, G., Zaldivar, D., Perez-Cisneros, M., Multilevel thresholding segmentation based on harmony search optimization, Journal of Applied Mathematics, 2013, 575414.
21. Ş. I. Birbil and S. C. Fang, "An electromagnetism-like mechanism for global optimization," *J. Glob. Optim.*, vol. 25, no. 1, pp. 263–282, 2003.
22. A. M. A. C. Rocha and E. M. G. P. Fernandes, "Modified movement force vector in an electromagnetism-like mechanism for global optimization," *Optim. Methods Softw.*, vol. 24, no. 2, pp. 253–270, 2009.
23. H. L. Hung and Y. F. Huang, "Peak to average power ratio reduction of multicarrier transmission systems using electromagnetism-like method," *Int. J. Innov. Comput. Inf. Control*, vol. 7, no. 5, pp. 2037–2050, 2011.
24. A. Yurtkuran and E. Emel, "A new Hybrid Electromagnetism-like Algorithm for capacitated vehicle routing problems," *Expert Syst. Appl.*, vol. 37, no. 4, pp. 3427–3433, 2010.
25. J. Y. Jhang and K. C. Lee, "Array pattern optimization using electromagnetism-like algorithm," *AEU - Int. J. Electron. Commun.*, vol. 63, pp. 491–496, 2009.
26. C. H. Lee and F. K. Chang, "Fractional-order PID controller optimization via improved electromagnetism-like algorithm," *Expert Syst. Appl.*, vol. 37, no. 12, pp. 8871–8878, 2010.
27. L. N. De Castro and F. J. Von Zuben, "Learning and optimization using the clonal selection principle," *IEEE Trans. Evol. Comput.*, vol. 6, no. 3, pp. 239–251, 2002.
28. A. M. A. C. Rocha and E. M. G. P. Fernandes, "Hybridizing the Electromagnetism-like algorithm with Descent Search for Solving Engineering Design Problems," *Int. J. Comput. Math.*, vol. 86, no. 10–11, pp. 1932–1946, 2009.
29. P. Ghamisi, M. S. Couceiro, J. A. Benediktsson, and N. M. F. Ferreira, "An efficient method for segmentation of images based on fractional calculus and natural selection," *Expert Syst. Appl.*, vol. 39, no. 16, pp. 12407–12417, 2012.
30. P. Wu, W.-H. Yang, and N.-C. Wei, "An Electromagnetism Algorithm of Neural Network Analysis—an Application To Textile Retail Operation," *J. Chinese Inst. Ind. Eng.*, vol. 21, no. 1, pp. 59–67, 2004.
31. K. De Jong, "Learning with genetic algorithms: An overview," *Mach. Learn.*, vol. 3, pp. 121–138, 1988.
32. B. Naderi, R. Tavakkoli-Moghaddam, and M. Khalili, "Electromagnetism-like mechanism and simulated annealing algorithms for flowshop scheduling problems minimizing the total weighted tardiness and makespan," *Knowledge-Based Syst.*, vol. 23, no. 2, pp. 77–85, 2010.
33. Z. Wang, A. C. Bovik, H. R. Sheikh, and E. P. Simoncelli, "Image quality assessment: From error visibility to structural similarity," *IEEE Trans. Image Process.*, vol. 13, no. 4, pp. 600–612, 2004.

34. D. Z. Lin Zhang, Lei Zhang, XuanqinMou, "FSIM : A Feature Similarity Index for Image," *IEEE Trans. Image Process.*, vol. 20, no. 8, pp. 2378–2386, 2011.
35. C. Tsallis, "Entropic nonextensivity: A possible measure of complexity," *Chaos, Solitons and Fractals*, vol. 13, pp. 371–391, 2002.
36. S. García, D. Molina, M. Lozano, and F. Herrera, "A Study on the Use of Non-Parametric Tests for Analyzing the Evolutionary Algorithms' Behaviour: A Case Study on the CEC 2005 Special Session on Real Parameter Optimization," *J. Heuristics*, vol. 15, pp. 617–644, 2009.

Chapter 3
Multi-circle Detection on Images

Abstract Hough transform (HT) represents the most common method for circle detection, exhibiting robustness and parallel processing. However, HT adversely demands a considerable computational load and large storage. Alternative approaches may include heuristic methods with iterative optimization procedures for detecting multiple circles. In this chapter a new circle detector for image processing is presented. In the approach, the detection process is therefore assumed as a multi-modal problem which allows multiple circle detection through only one optimization procedure. The algorithm uses a combination of three non-collinear edge points as parameters to determine circles candidates. A matching function (nectar amount) determines if such circle candidates (bee-food-sources) are actually present in the image. Guided by the values of such matching function, the set of encoded candidate circles are evolved through the Artificial Bee Colony (ABC) algorithm so the best candidate (global optimum) can be fitted into an actual circle within the edge-only image. An analysis of the incorporated exhausted-sources memory is executed in order to identify potential local optima i.e. other circles. The overall approach yields a fast multiple-circle detector that locates circular shapes delivering sub-pixel accuracy despite complicated conditions such as partial occluded circles, arc segments or noisy images.

3.1 Introduction

The problem of detecting circular features holds paramount importance for image analysis in industrial applications such as automatic inspection of manufactured products and components, aided vectorization of drawings, target detection, etc. [1]. Solving common challenges for object localization is normally approached by two techniques. First, all deterministic techniques including the application of Hough transform-based methods [2], the use of geometric hashing and other template or model-based matching techniques [3, 4]. All stochastic techniques are gathered inside a second set including random sample consensus techniques [5], simulated annealing [6] and optimization algorithms [7–9].

© Springer International Publishing AG 2017
E. Cuevas et al., *Evolutionary Computation Techniques:*
A Comparative Perspective, Studies in Computational Intelligence 686,
DOI 10.1007/978-3-319-51109-2_3

Template and model matching techniques have been the first approaches to be applied to shape detection. Although several methods have been developed for solving such a problem [10], shape coding techniques and combination of shape properties have been successfully employed for representing different objects. Their main drawback is related to the contour extraction step from real images and their deficiencies to deal with pose invariance except for very simple objects.

The circle detection in digital images is commonly solved through the Circular Hough Transform [11]. A typical Hough-based approach employs an edge detector and some edge information to infer locations and radii values. Peak detection is then performed by averaging, filtering and histogramming within the transform space. Unfortunately, such approach requires a large storage space as the 3-D cells include parameters (x, y, r) which augments the computational complexity and yields a low processing speed. The accuracy of the extracted parameters for the detected-circle is poor, particularly in presence of noise [12].

In the particular case of a digital image holding a significant width and height and some densely populated edge pixels, the required processing time for Circular Hough Transform makes it prohibitive to be deployed in real time applications. In order to overcome such a problem, some other researchers have proposed new approaches following Hough transform principles, yielding the probabilistic Hough transform [13], the randomized Hough transform (RHT) [14], the fuzzy Hough transform [15] and some other topics presented by Becker in [16].

As an alternative to Hough Transform-based techniques, the problem of shape recognition in computer vision has also been handled through optimization methods. Ayala-Ramirez et al. presented a Genetic Algorithm (GA) based circle detector [17] which is capable of detecting multiple circles over real images but fails frequently while detecting imperfect circles. On the other hand, Dasgupta et al. [18] has recently proposed an automatic circle detector using the Bacterial Foraging Optimization Algorithm (BFOA) as optimization procedure. However, both methods employ an iterative scheme to achieve multiple circle detection which executes the algorithm as many times as the number of circles to be found. Only one circle can be found at each run yielding quite long execution times.

An impressive growth in the field of biologically inspired meta-heuristics for search and optimization has emerged during the last decade. Some bio-inspired examples like Genetic Algorithm (GA) [19] and Differential Evolution (DE) [20] have been applied to solve complex optimization problems, while Swarm Intelligence (SI) has recently attracted interest from several fields. The SI core lies on the analysis of collective behavior of relatively simple agents working on decentralized systems. Such systems typically gather an agent's population that can communicate to each other while sharing a common environment. Despite a non-centralized control algorithm regulates the agent behavior, the agent can solve complex tasks by analyzing a given global model and harvesting cooperation to other agents. Therefore, a novel global behavior evolves from interaction among agents as it can be seen on typical examples that include ant colonies, animal herding, bird flocking, fish schooling, honey bees, bacteria, and many more.

Swarm-based algorithms, such as Particle Swarm Optimization [21], Ant Colony Optimization [22] and Bacterial Foraging Optimization Algorithm (BFOA) [23] have already been successfully applied to solve several engineering applications.

Karaboga has recently presented one bee-swarm algorithm for solving numerical optimization problems known as the Artificial Bee Colony (ABC) method [24]. ABC is a population-based algorithm which mimics the intelligent foraging behavior of a honey-bee swarm. The food source's location represents a candidate solution for the optimization problem while the corresponding nectar availability accounts for quality i.e. the fitness objective function of the associated solution. The ABC method has already been successfully applied to solve different sorts of engineering challenges at several fields such as signal processing [25] and electromagnetism [26]. However, to the best of our knowledge, ABC has not been applied to any image processing task until date.

First, ABC generates a randomly distributed initial population (food source locations). After initialization, an objective function evaluates whether such candidates represent an acceptable solution (nectar amount) or not. Guided by the values of such objective function, the candidate solutions are evolved through different ABC operators. When the fitness function (nectar amount) cannot be further improved after reaching the maximum number of cycles, its related food source is assumed to be abandoned and replaced by a new randomly chosen food source location. However, in order to contribute towards the solution of multi-modal optimization problems, our proposal considers that such abandoned solutions are not to be discarded at all, but they are to be arranged into a so-called "exhausted-sources memory" which contains valuable information regarding global and local optima that have been emerging as the optimization evolves.

Although ABC draws several similarities to other bio-inspired algorithms, there are some significant issues to be discussed: ABC does not depend upon the best member within the population in order to update the particle's motion as it is done by PSO [27], neither it does not require all particles for computing parameters, such as the pheromone concentration, which determines the overall performance as it is demanded by ACO [28]. For ABC, each particle randomly supports the definition of new motion vectors, contributing towards augmenting the population diversity. Similarly to DE, ABC does require a selection operation that allows individuals to access a fair chance of being elected for recombination (diversity). However, ABC holds a second modification operation which follows a random "roulette selection" allowing some privileges for best located solutions and augmenting the convergence speed [29]. In contrast to the local particle modifications executed by BFOA, ABC employs some operators that tolerate modifications over the full search space for each parameter, avoiding typical oscillations in contrast to BFOA which operates over multimodal surfaces [30]. Previous works in [31, 32] have shown that the performance of ABC surpasses others algorithms such as the PSO and the DE algorithms.

This chapter presents an algorithm for the automatic detection of multiple circular shapes from complicated and noisy images without considering conventional

Hough transform principles. The detection process is approached as a multi-modal optimization problem. The ABC algorithm searches the entire edge-map looking for circular shapes by using the combination of three non-collinear edge points that represent candidate circles (food source locations) in the edge-only image of the scene. An objective function is used to measure the existence of a candidate circle over the edge map. Guided by the values of such objective function, the set of encoded candidate circles are evolved through the ABC algorithm so that the best candidate can be fitted into the most circular shape within the edge-only image. A subsequent analysis of the incorporated exhausted-source memory is then executed in order to identify potential useful local optima (other circles). The approach generates a fast sub-pixel detector which can effectively identify multiple circles in real images despite circular objects exhibiting significant occluded sections. Experimental evidence shows the effectiveness of the method for detecting circles under various conditions. A comparison to one state-of-the-art GA-based method [17], the BFOA [18] and the RHT algorithm [14] on different images has been included to demonstrate the performance of the proposed approach. Conclusions of the experimental comparison are validated through statistical tests that properly support the discussion.

The chapter is organized as follows: Sect. 3.2 provides information regarding the ABC algorithm. Section 3.3 depicts the implementation of the proposed circle detector. The complete multiple circle detection procedure is presented by Sect. 3.4. Experimental results for the proposed approach are stated in Sect. 3.5 and some relevant conclusions are discussed in Sect. 3.6.

3.2 Artificial Bee Colony (ABC) Algorithm

3.2.1 Real Bee Profile

Tasks are performed by specialized individuals in real bee colonies. Such highly specialized bees seek to maximize the nectar amount stored in the hive by using an efficient labour assignment and self-organization. The minimal model for a honey bee colony consists of three kinds: employed bees, onlooker bees and scout bees. Half of the colony consists of employed bees while the other half includes onlooker bees. However, either employed or onlooker may become scout bees if certain conditions are met. Employed bees are responsible for exploiting the nectar sources explored before and giving information to the waiting bees (onlooker bees) working at the hive. Such information refers to the quality of the food source sites which they are exploiting. Onlooker bees wait in the hive and decide on whether to exploit a given food source according to the employed bees report. Scouts randomly search the environment in order to find new food sources depending on internal motivation.

The bee's behaviour can be summarized as follows:

1. At the initial phase of the foraging process, bees start to randomly explore the environment in order to find a food source.
2. After a food source is found, the bee becomes an employed bee and starts exploiting the discovered source. The employed bee returns to the hive with nectar and unloads it. The bee can go back to its discovered site directly or it can share information about its source by performing a dance over the dancing area. Once its source is exhausted, it becomes a scout and starts a random search for new sources.
3. Onlooker bees, waiting at the hive, watch the dances advertising the profitable sources and choose the next source considering the frequency of dancing which is proportional to the quality of the source [30].

3.2.2 Description of the ABC Algorithm

The Artificial Bee Colony (ABC) algorithm, proposed by Karaboga in 2005 for real-parameter optimization, is a recently introduced optimization algorithm which simulates the foraging behaviour of bees working within a colony. ABC provides a population-based search procedure under which all individuals, also known as food positions, are defined within the multidimensional space. They are available for modification by artificial bees through the time. Therefore, a bee aims to discover the location of food sources showing higher nectar amounts (better solutions), choosing the one which holds the highest record. If a food source is exhausted during the foraging process, it is then abandoned while a scout bee is expected to discover a new one. A simple analogy between main procedures of the ABC and the natural bee colony are shown below.

3.2.3 Initializing the Population

The algorithm begins by initializing N_p food sources for the employed bees. In nature, the bee's hive will be surrounded by potential food sources that need to be discovered, evaluated and harvested. Within the ABC algorithm framework, potential food sources are contained within the search space with initial candidate solutions (initial food source locations). Each food source is a D-dimensional vector containing the parameter values to be optimized which are randomly and uniformly distributed between the pre-specified lower initial parameter bound x_j^{low} and the upper initial parameter bound x_j^{high}.

$$x_{j,i} = x_j^{low} + \text{rand}(0,1) \cdot (x_j^{high} - x_j^{low});$$
$$j = 1, 2, \ldots, D; \quad i = 1, 2, \ldots, N_p. \tag{3.1}$$

with j and i being the parameter and individual indexes respectively. Hence, $x_{j,i}$ is the jth parameter of the ith individual. For each food source i, a counter A_i is assigned and incremented as the food source's fitness does not improve for the current evaluation. If a given counter A_i reaches a "*limit*" number, then the food source i is considered to be exhausted and a new food source location is required. Once initial food sources have been generated, bees cooperate to find the best possible solution. The final solution is considered after a pre-defined cycle number is reached.

3.2.4 Sending Employed Bees to Food Sources

Each employed bee is associated to only one food source location. The number of food source sites is equal to the number of employed bees. An employed bee will produce a location modification for the food source under consideration. Such variations depend on the positions of other food sources. The employed bee will consider only one of the food sources in order to evaluate a new location (food source position). Variations on food sources locations within the ABC algorithm are performed via the next expression:

$$v_{j,i} = x_{j,i} + \phi_{j,i}(x_{j,i} - x_{j,k});$$
$$k \in \{1, 2, \ldots, N_p\}; \quad j \in \{1, 2, \ldots, D\} \tag{3.2}$$

$x_{j,i}$ is a randomly chosen j parameter of the ith individual and k is one of the N_p food sources, satisfying the conditional $i \neq k$. If a given parameter of the candidate solution v_i exceeds its predetermined boundaries, that parameter should be adjusted in order to fit the appropriate range. The scale factor $\phi_{j,i}$ is a random number between $[-1, 1]$. Once the new solution is generated, a fitness value representing the profitability associated to a particular solution is calculated. The fitness value for a minimization problem can be assigned to a solution v_i by the following expression

$$fit_i = \begin{cases} \dfrac{1}{1+J_i} & \text{if } J_i \geq 0 \\ 1 + abs(J_i) & \text{if } J_i < 0 \end{cases} \tag{3.3}$$

where J_i is the objective function to be minimized.

A greedy selection process is thus applied between v_i and x_i. If the nectar amount (fitness) of v_i is better, then the solution x_i is replaced by v_i and A_i is cleared, otherwise x_i remains and A_i is incremented with a value of 1.

3.2.5 Calculate Probability Values of Food Source Positions

In a bee colony, bees share information about the nectar associated to food sources by dancing. The onlooker evaluates the dancing of its nest mates and decides whether to go to a determined food source or not, depending on its profitability. Inside the ABC's framework, such behaviour is simulated by a probabilistic selection tracking the fitness value fit_i of each solution in the population which holds an associated probability depending on its fitness. Such probability can be calculated by the following expression:

$$Prob_i = \frac{fit_i}{\sum_{i=1}^{N_p} fit_i} \tag{3.4}$$

3.2.6 Sending Onlooker Bees to Seek for Food Source Sites

At this stage, by using the probability computed in Eq. (3.4), it is possible to simulate the transformation of onlooker bees into employee bees. The onlooker bee number in ABC is N_p. The onlooker bee chooses the food source to exploit via a probabilistic selection. In the basic ABC algorithm a roulette wheel selection scheme is used, i.e. a random real number between [0 1] is assigned to each food source. If the probability $Prob_i$ for each food source location is greater than the random number, the onlooker generates a variation for that food source as in Eq. (3.2). At this point, the onlooker bee is behaving as an employed bee, thus the same process is applied for the evaluation of the newly generated food sources, i.e. a greedy selection is performed between both the original and the mutant solution. In case the nectar amount i.e. fitness of the new solution is better than before, such position is held, the counter A_i is cleared and the old one is forgotten. Otherwise the old solution remains and its counter is increased by one. The process is repeated until all onlooker bees had attended a food source. As this is a probabilistic selection process, food source locations holding most nectar, i.e. the highest probability, are more likely to be chosen by the onlooker bees.

3.2.7 Abandoning Exhausted Food Sources

At each ABC cycle, any exhausted food source must be detected. A food source is considered to be exhausted once its associated counter A_i reaches the "*limit*" value, which is a control parameter of the ABC that should be specified before the algorithm iterates. Once a food source is marked as exhausted, it must be replaced with a new one which must be provided by a scout bee. A scout bee explores the searching space with no previous information i.e. the new solution is generated randomly as in Eq. (3.1). The basic ABC algorithm considers a maximum of one exhausted food source per cycle. Considering any food source might be visited by a bee, the new solution has to be evaluated to verify its profitability according to Eq. (3.3). A new food source is now available to be exploited by employed bees. However, in order to solve multi-modal optimization problems, exhausted food

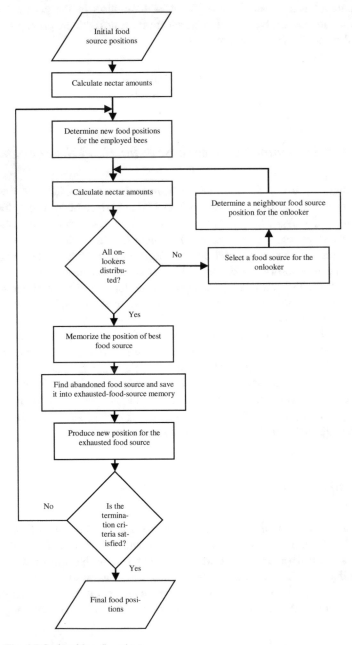

Fig. 3.1 The ABC algorithm flowchart

sources (solutions) are not to be discarded but stored at the exhausted-source memory as they contain valuable information regarding global and local optima which has emerged during the optimization process.

Figure 3.1 shows the flowchart of the ABC algorithm as it has been stated at this section.

3.3 Circle Detection Using ABC

At this work, circles are represented by a well-known second degree equation (see Eq. 3.5) that passes through three points in the edge map. Pre-processing includes a classical Canny edge detector which uses a single-pixel contour marker and stores the location for each edge point. Such points are the only potential candidates to define circles by considering triplets. All the edge points are thus stored within a vector array $P = \{p_1, p_2, \ldots, p_{E_p}\}$ with E_p being the total number of edge pixels in the image. The algorithm saves the (x_i, y_i) coordinates for each edge pixel p_i within the edge vector.

In order to construct each circle candidate (or food positions within the ABC-framework), the indexes i_1, i_2 and i_3 of three non-collinear edge points must be combined, assuming that the circle's contour goes through points p_{i_1}; p_{i_2}; p_{i_3}. A number of candidate solutions are generated randomly for the initial set of food sources. Solutions will thus evolve through the application of the ABC upon food sources until a minimum is reached, considering the best individual as the solution for the circle detection problem. A further analysis searches for other circular patterns that may be partially marked during the detection process.

Applying the classic Hough Transform for circle detection would normally require huge amounts of memory and consume large computation time. In order to reach a sub-pixel resolution (just like the method discussed at this chapter), classic Hough Transform methods also consider three edge points to cast a vote for the corresponding point within the parameter space. An evidence-collecting step is also required. As the overall evolution process continues, the objective function improves at each generation by discriminating non-plausible circular shapes, locating circles and avoiding useless visits to several other image points. An overview of the required steps to formulate the circle detection task as an ABC optimization problem is presented below.

3.3.1 Individual Representation

Each candidate solution C (food source) uses three edge points. Under such representation, edge points are stored following a relative positional index within the edge array P. In turn, the procedure will encode a food source as the circle that passes through three points p_i, p_j and p_k ($C = \{p_i, p_j, p_k\}$). Each circle C is thus represented by three parameters x_0, y_0 and r, being (x_0, y_0) the (x, y) coordinates of the centre of the circle and r its radius. The equation of the circle passing through the three edge points can thus be computed as follows:

$$(x - x_0)^2 + (y - y_0)^2 = r^2 \tag{3.5}$$

considering

$$\mathbf{A} = \begin{bmatrix} x_j^2 + y_j^2 - (x_i^2 + y_i^2) & 2 \cdot (y_j - y_i) \\ x_k^2 + y_k^2 - (x_i^2 + y_i^2) & 2 \cdot (y_k - y_i) \end{bmatrix}$$

$$\mathbf{B} = \begin{bmatrix} 2 \cdot (x_j - x_i) & x_j^2 + y_j^2 - (x_i^2 + y_i^2) \\ 2 \cdot (x_k - x_i) & x_k^2 + y_k^2 - (x_i^2 + y_i^2) \end{bmatrix}, \tag{3.6}$$

$$x_0 = \frac{\det(\mathbf{A})}{4((x_j - x_i)(y_k - y_i) - (x_k - x_i)(y_j - y_i))},$$

$$y_0 = \frac{\det(\mathbf{B})}{4((x_j - x_i)(y_k - y_i) - (x_k - x_i)(y_j - y_i))}, \tag{3.7}$$

and

$$r = \sqrt{(x_0 - x_d)^2 + (y_0 - y_d)^2}, \tag{3.8}$$

being $\det(.)$ the determinant and $d \in \{i, j, k\}$. Figure 3.2 illustrates the parameters defined by Eqs. 3.5–3.8.

Therefore it is possible to represent the shape parameters (for the circle, $[x_0, y_0, r]$) as a transformation T of the edge vector indexes i, j and k.

$$[x_0, y_0, r] = T(i, j, k) \tag{3.9}$$

with T being the transformation calculated after the previous computations of x_0, y_0, and r. By exploring each index as an individual parameter, it is feasible to apply the

Fig. 3.2 Circle candidate (individual) built from the combination of points p_i, p_j and p_k

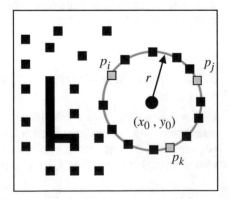

ABC algorithm for seeking appropriate circular parameters. The approach reduces the search space by eliminating unfeasible solutions.

3.3.2 Objective Function

Optimization refers to choosing the best element from one set of available alternatives. In the simplest case, it means to minimize an objective function or error by systematically choosing the values of variables from their valid ranges. In order to calculate the error produced by a candidate solution C, the circumference coordinates are calculated as a virtual shape which, in turn, must also be validated, i.e. if it really exists in the edge image. The test set is represented by $S = \{s_1, s_2, \ldots, s_{N_s}\}$, where N_s are the number of points over which the existence of an edge point, corresponding to C, should be tested.

The set S is generated by the midpoint circle algorithm [33]. The Midpoint Circle Algorithm (MCA) is a searching method which seeks required points for drawing a circle. Any point (x, y) on the boundary of the circle with radius r satisfies the equation $f_{Circle}(x, y) = x^2 + y^2 - r^2$. However, MCA avoids computing square-root calculations by comparing the pixel separation distances. A method for direct distance comparison is to test the halfway position between two pixels (sub-pixel distance) to determine if this midpoint is inside or outside the circle boundary. If the point is in the interior of the circle, the circle function is negative. Thus, if the point is outside the circle, the circle function is positive. Therefore, the error involved in locating pixel positions using the midpoint test is limited to one-half the pixel separation (sub-pixel precision). To summarize, the relative position of any point (x, y) can be determined by checking the sign of the circle function:

$$f_{Circle}(x, y) \begin{cases} < 0 & \text{if } (x, y) \text{ is inside the circle boundary} \\ = 0 & \text{if } (x, y) \text{ is on the circle boundary} \\ > 0 & \text{if } (x, y) \text{ is outside the circle boundary} \end{cases} \quad (3.10)$$

The circle-function test in Eq. 3.10 is applied to mid-positions between pixels nearby the circle path at each sampling step. Figure 3.3a shows the midpoint between the two candidate pixels at sampling position x_k.

In MCA the computation time is reduced by considering the symmetry of circles. Circle sections in adjacent octants within one quadrant are symmetric with respect to the 45° line dividing the two octants. These symmetry conditions are illustrated in Fig. 3.3b, where a point at position (x, y) on a one-eighth circle sector is mapped into the seven circle points in the other octants of the xy plane. Taking advantage of the circle symmetry, it is possible to generate all pixel positions around a circle by calculating only the points within the sector from $x = 0$ to $x = y$. Thus, in this chapter, the MCA is used to calculate the required S points that represent the circle candidate C. The algorithm can be considered the quickest providing a sub-pixel

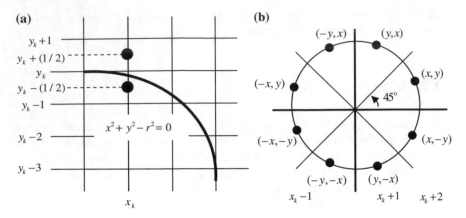

Fig. 3.3 **a** Symmetry of a circle: calculation of a circle point (x, y) in one octant yields the circle points shown for other seven octants. **b** Midpoint between candidate pixels at sampling position x_k along a circular path

precision [34]. However, in order to protect the MCA operation, it is important to assure that points lying outside the image plane must not be considered in S.

The objective function $J(C)$ represents the matching error produced between the pixels S of the circle candidate C (food source) and the pixels that actually exist in the edge image, yielding:

$$J(C) = 1 - \frac{\sum_{v=1}^{Ns} E(x_v, y_v)}{Ns} \tag{3.11}$$

where $E(x_i, y_i)$ is a function that verifies the pixel existence in (x_v, y_v), with $(x_v, y_v) \in S$ and N_s being the number of pixels lying on the perimeter corresponding to C currently under testing. Hence, function $E(x_v, y_v)$ is defined as:

$$E(x_v, y_v) = \begin{cases} 1 & \text{if the pixel } (x_v, y_v) \text{ is an edge point} \\ 0 & \text{otherwise} \end{cases} \tag{3.12}$$

A value near to zero of $J(C)$ implies a better response from the "circularity" operator. Figure 3.4 shows the procedure to evaluate a candidate solution C with its representation as a virtual shape S. Figure 3.4b, the virtual shape is compared to the original image, point by point, in order to find coincidences between virtual and edge points. The virtual shape is built from points p_i, p_j and p_k shown by Fig. 3.4a. The virtual shape S gathers 56 points ($Ns = 56$) with only 18 of such points existing in both images (shown as blue points plus red points in Fig. 3.4c) yielding: $\sum_{v=1}^{Ns} E(x_v, y_v) = 18$ and therefore $J(C) \approx 0.67$.

Typically two stop criteria have been employed for meta-heuristic algorithms: either an upper limit of the fitness function is reached or an upper limit of the

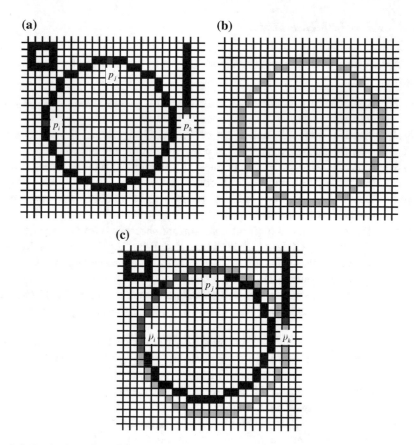

Fig. 3.4 Evaluation of candidate solutions **C**. The image in **a** shows the original image while **b** presents the virtual shape generated including points p_i, p_j and p_k. The image in **c** shows coincidences between both images marked by *blue* or *red* pixels while the virtual shape is also depicted in *green*

number of generations is attained [35]. The first criterion requires an extensive knowledge of the problem and its feasible solutions [36]. On the contrary, by considering the stop criterion based on the number of generation, feasible solutions may be found by exhaustive exploration of the search space. At this chapter, the number of iterations as stop criterion is employed in order to allow the multi-circle detection. Hence, if a solution representing a valid circle appears at early steps, it would be stored at the exhausted-source memory and the algorithm continues detecting other feasible solution until the number of iterations is reached. Therefore, the main issue is to define a fair iteration number which should be big enough as to allow finding circles at the image and small enough as to avoid an exaggerated computational cost. In this chapter, such number was experimentally defined to 300.

3.3.3   ABC Implementation

The implementation of the proposed algorithm can be summarized by the following steps:

Step 1   Apply the Canny filter to the original image and store the edge pixels within the vector P.

Step 2   Initialize required parameters of the ABC algorithm. Set the colony's size, the abandonment *limit* and the maximum number of cycles.

Step 3   Initialize N_C circle candidates C_b (original food sources) where $b \in (1, \ldots, N_C)$, and clear all counters A_b.

Step 4   Obtain the matching fitness (food source quality) for each circle candidate C_b using Eq. 3.10.

Step 5   Repeat steps 6–10 until a maximum number of cycles are reached.

Step 6   Modify the circle candidates as stated by Eq. 3.2 and evaluate its matching fitness (send employed bees onto food sources). Likewise, update all counters A_b.

Step 7   Calculate the probability value $Prob_b$ for each circle candidate C_b. Such probability value will be used as a preference index by onlooker bees (Eq. 3.4).

Step 8   Generate new circles candidates (using the Eq. 3.2) from current candidates according to their probability $Prob_b$ (Send onlooker bees to their selected food source). Likewise, update counters A_b.

Step 9   Obtain the matching fitness for each circle candidate C_b and calculate the best circle candidate (solution).

Step 10   Stop modifying the circle candidate C_b (food source) whose counter A_b has reached its counter "*limit*" and save it as a possible solution (global or local optimum) in the exhausted-source memory. Clear A_b and generate a new circle candidate according to Eq. 3.1.

Step 11   Analyze solutions previously stored on the exhausted-source memory (see Sect. 3.5). The memory holds solutions (any other potential circular shape in the image) generated through the evolution of the optimization algorithm.

3.4   The Multiple Circle Detection Procedure

The original ABC algorithm considers the so-called abandonment limit which aims to stop the search for a feasible solution after a trial number is reached. All "stuck solutions", i.e. those that do not improve further during the optimization cycle are supposed to be discarded and replaced by other randomly generated solutions. However, this chapter proposes the use of an "exhausted-source memory" to store information regarding local optima which represent feasible solutions for the multi-circle detection problem.

Several heuristic methods have been employed for detecting multiple circles as an alternative to classical Hough transform-based techniques [17, 18]. Such strategies imply that only one circle can be marked per optimization cycle, forcing a multiple execution of the algorithm in order to achieve multiple circle detection. The surface representing $J(C)$ holds a multimodal nature which contains several global and local optima that are related to potential circular shapes in the edge map. Therefore, this chapter aims to solve the objective function $J(C)$ using only one optimization procedure by assuming the multi-detection problem as a multimodal optimization issue. Guided by the values of matching function, the set of encoded circle candidates are evolved through the ABC algorithm and the best circle candidate (global optimum) is considered to be the first detected circle over the edge-only image.

It is important to consider that many local optima may represent the same circle. In order to differentiate between the true potential circle and any other circular shapes (i.e. other local minima held at exhausted-source memory), a distinctiveness factor $E_{S_{di}}$ is used to measure the mismatch between two circles as follows:

$$E_{S_{di}} = \sqrt{(x_A - x_B)^2 + (y_A - y_B)^2 + (r_A - r_B)^2} \tag{3.13}$$

being (x_A, y_A) and r_A central coordinates and radius of the circle C_A respectively, while (x_B, y_B) and r_B are the corresponding parameters of the circle C_B. In order to decide whether two circles must be considered different, a threshold value $E_{S_{TH}}$ can be defined as follows:

$$E_{S_{TH}} = \alpha \sqrt{(cols - 1)^2 + (rows - 1)^2 + (r_{max} - r_{min})^2} \tag{3.14}$$

where *rows* and *cols* refer to the number of rows and columns of the image respectively, r_{max} and r_{min} are the maximum and minimum radii for representing feasible candidate circles in the image, while α is a sensitivity factor affecting the discrimination between the circles. A high value of α allows circles to be significantly different and still be considered as the same while a low level would imply that two circles with slight differences in radius or position could be considered as different. The $E_{S_{TH}}$ value calculated by Eq. 3.13 allows discriminating circles with no consideration of the image size.

In order to find "sufficiently different" circles, the exhausted-source memory is sorted decreasingly according to its matching fitness $J(C)$. A solution would be considered good enough if its distinctiveness factor $E_{S_{di}}$ surpasses a threshold $E_{S_{TH}}$.

The multiple circle detection procedure can thus be summarized as follows:

Step 1 The best solution found by the ABC algorithm and all candidate circles held by the exhausted-source memory are organized decreasingly with respect to their matching fitness, yielding a new vector $M_C = \{C_1, \ldots, C_{Ne}\}$, with Ne being the size of the exhausted-source memory plus 1.

Step 2 The candidate circle C_1 showing the highest matching fitness is identified as the first circle shape CS_1 as it is stored within a vector Ac of actual circles.

Step 3 The distinctiveness factor $E_{s_{di}}$ for the candidate circle C_m (element m in M_C) is compared to every element in Ac. If $E_{s_{di}} > E_{S_{TH}}$ is true for each pair of solutions (those present in Ac and the candidate circle C_m), then C_m is considered as a new circle shape CS and it is added to the vector Ac. Otherwise, the next circle candidate C_{m+1} is evaluated and C_m is discarded.

Step 4 Step 3 is repeated until all Ne candidate circles in M_C have been analyzed.

Summarizing the overall procedure, Fig. 3.5 shows the outcome of the multi-modal optimization search. The input image (Fig. 3.5a) has a resolution of 256×256 pixels and shows two circles and two ellipses with a different circularity factor. Figure 3.5b presents the detected circles with a green overlay. Figure 3.5c shows the candidate circles held by the exhausted-source memory after the optimization process while Fig. 3.5d presents the results after the discrimination procedure described in this section.

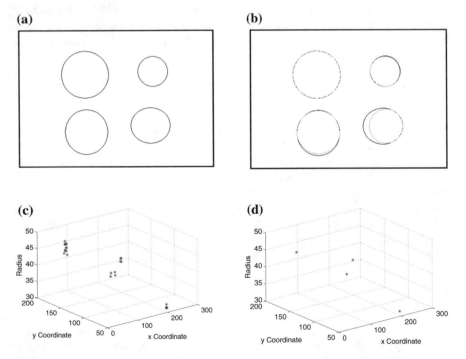

Fig. 3.5 Performance of the ABC based circular detector showing: **a** the original image and **b** all detected circles as an overlay. **c** Candidate circles held by the exhausted-source memory after the optimization process and **d** all detected circles after the discrimination process

3.5 Experimental Results

Experimental tests have been developed in order to evaluate the performance of the circle detector. The experiments address the following tasks:

(1) Circle localization,
(2) Shape discrimination,
(3) Circular approximation: occluded circles and arc detection.

Table 3.1 presents the parameters for the ABC algorithm at this work. They have been kept for all test images after being experimentally defined.

All the experiments have been executed over a Pentium IV 2.5 GHz computer under C language programming. All the images are pre-processed by the standard Canny edge-detector using the image-processing toolbox for MATLAB R2008a.

3.5.1 Circle Localization

3.5.1.1 Synthetic Images

The experimental setup includes the use of several synthetic images of 320 × 240 pixels. All images contain a different amount of circular shapes and some have also been contaminated by added noise as to increase the complexity of the localization task. The algorithm is executed over 100 times for each test image, successfully identifying and marking all required circles in the image. The detection has proved to be robust to translation and scaling still offering a reasonably low execution time. Figure 3.6 shows the outcome after applying the algorithm to two images from the experimental set.

3.5.1.2 Natural Images

This experiment tests the circle detection upon real-life images. All twenty five test images of 640 × 480 pixels have been captured using a digital camera under an 8-bit colour format. Each natural scene includes circular shapes which have been

Table 3.1 ABC detector parameters

Colony size	Abandonment limit	Number of cycles	α	Limit
20	100	300	0.05	30

(a)   (b)

(c)   (d)

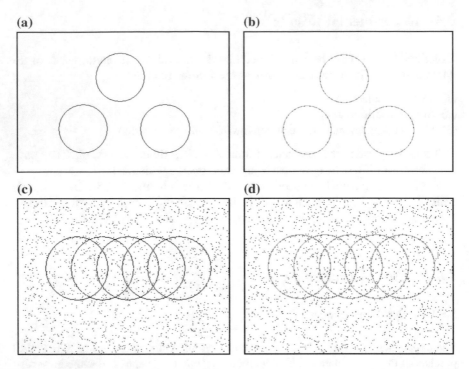

Fig. 3.6 Circle localization over synthetic images. The image **a** shows the original image while **b** presents the detected circles as an overlay. The image in **c** shows a second image with salt and pepper noise and **d** shows detected circles as a *green* overlay

pre-processed using the Canny edge detection algorithm before being fed to the ABC procedure. The example in Fig. 3.7 focuses at a relevant issue as follows: Fig. 3.7c presents some candidate circles which have been stored at the exhausted-source memory, just after the optimization process has finished. Figure 3.7d shows the results after applying the discrimination procedure from Sect. 3.4.

3.5.2   Shape Discrimination Tests

This section discusses on the detector's ability to differentiate circular patterns over any other shape which might be present in the image. Figure 3.8 shows five synthetic images of 540×300 pixels with added noise portraying a number of different shapes. Figure 3.9 repeats the experiment over real-life images.

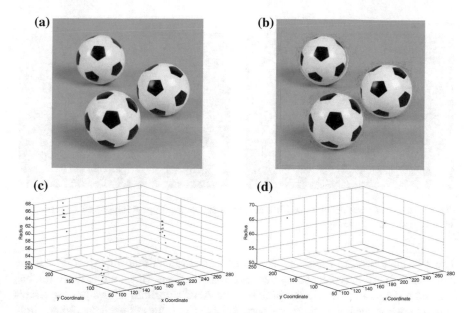

Fig. 3.7 Circle detection algorithm over natural images: the image in **a** shows the original image while **b** presents the detected circles as a *green* overlay. The image in **c** represents candidate circles lying at the exhausted-source memory after the optimization has finished and **d** presents all detected circles just after finishing the discrimination process

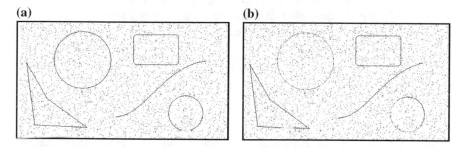

Fig. 3.8 Shape discrimination over synthetic images: **a** shows the original image contaminated by salt and pepper noise while **b** presents the detected circles as an overlay

3.5.3   Circular Approximation: Occluded Circles and Arc Detection

In this chapter, the circle detection algorithm may also be useful to approximate circular shapes from arc segments, occluded circular shapes or imperfect circles.

(a)   (b)

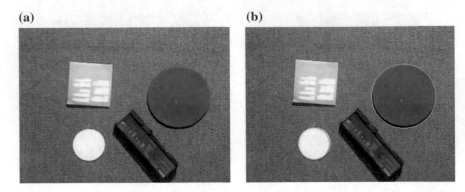

Fig. 3.9 Shape discrimination in real-life images: **a** shows the original image and **b** presents the detected circle as an overlay

The relevance of such functionality comes from the fact that imperfect circles are commonly found in typical computer vision applications. Since circle detection has been considered an optimization problem, the ABC algorithm allows finding circles which may approach a given shape according to the values of each candidate circle just after evaluating the objective function $J(C)$. Figure 3.10a shows some examples of circular approximation. Likewise, the proposed algorithm is able to find circle parameters that better approach to an arc or an occluded circle. Figure 3.10b and 3.10c show some examples of this functionality. A small value for $J(C)$, i.e. near zero, refers to a circle while a slightly bigger value in C accounts for an arc or an occluded circular shape. Such fact does not represent any trouble as circles can be shown following the obtained $J(C)$ values.

3.5.4 Performance Evaluation

In order to enhance the algorithm analysis, the ABC proposed algorithm is compared to the BFAOA and the GA circle detectors over a set of common images.

The GA algorithm follows the proposal of Ayala-Ramirez et al., which considers the population size as 70, the crossover probability as 0.55, the mutation probability as 0.10 and the number of elite individuals as 2. The roulette wheel selection and the 1-point crossover operator are both applied. The parameter setup and the fitness function follow the configuration suggested in [17]. The BFAOA algorithm follows the implementation from [18] considering the experimental parameters as: $S = 50$, $N_c = 100$, $N_s = 4$, $N_{ed} = 1$, $P_{ed} = 0.25$, $d_{attract} = 0.1$, $w_{attract} = 0.2$, $w_{repellant} = 10$, $h_{repellant} = 0.1$, $\lambda = 400$ and $\psi = 6$. Such values are found as the best configuration set according to [18].

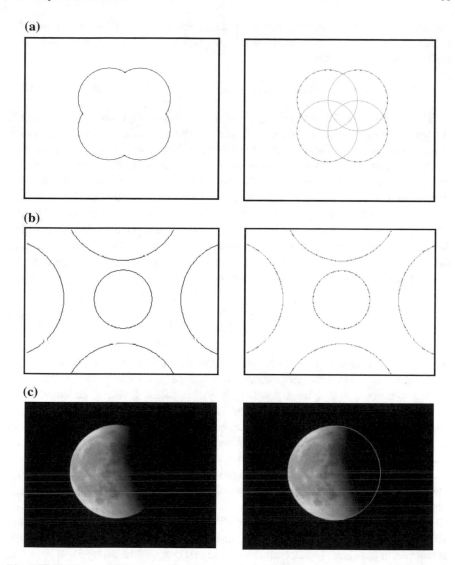

Fig. 3.10 ABC approximating circular shapes and arc sections

Images rarely contain perfectly-shaped circles. Therefore, with the purpose of testing accuracy for a single-circle, the detection is challenged by a ground-truth circle which is manually determined from the original edge-map. The parameters $(x_{true}, y_{true}, r_{true})$ of such testing circle are computed using the Eqs. 3.6–3.9 for three circumference points over the manually determined circle. If the centre and the radius of the detected circle are defined as (x_D, y_D) and r_D, then an Error Score (Es) can be chosen as follows:

$$\text{Es} = \eta \cdot (|x_{true} - x_D| + |y_{true} - y_D|) + \mu \cdot |r_{true} - r_D| \qquad (3.15)$$

The central point difference $(|x_{true} - x_D| + |y_{true} - y_D|)$ represents the centre shift for the detected circle as it is compared to a benchmark circle. The radio mismatch $(|r_{true} - r_D|)$ accounts for the difference between their radii. η and μ represent two weighting parameters which are to be applied separately to the central point difference and to the radio mismatch for the final error **Es**. At this work, they are chosen as $\eta = 0.05$ and $\mu = 0.1$. Such particular choice ensures that the radii difference would be strongly weighted in comparison to the difference of central circular positions between the manually detected and the machine-detected circles. In order to use an error metric for multiple-circle detection, the averaged **Es** produced from each circle in the image is considered. Such criterion, defined as the Multiple Error (ME), is calculated as follows:

$$\text{ME} = \left(\frac{1}{NC}\right) \cdot \sum_{R=1}^{NC} \text{Es}_R \qquad (3.16)$$

where **NC** represents the number of circles within the image. In case ME is less than 1, the algorithm gets a success; otherwise it has failed on detecting the circle set. Notice that for $\eta = 0.05$ and $\mu = 0.1$, it yields ME < 1 which accounts for a maximal tolerated average difference on radius length of 10 pixels, whereas the maximum average mismatch for the centre location can be up to 20 pixels. In general, the success rate (SR) can thus be defined as the percentage of reaching success after a certain number of trials.

Figure 3.11 shows three synthetic images and the resulting images after applying the GA-based algorithm [17], the BFOA method [18] and the proposed approach. Figure 3.12 presents experimental results considering three natural images. The performance is analyzed by considering 35 different executions for each algorithm. Table 3.2 shows the averaged execution time, the success rate in percentage and the averaged multiple error (ME), considering six test images (shown by Figs. 3.11 and 3.12). The best entries are bold-cased in Table 3.2. Close inspection reveals that the proposed method is able to achieve the highest success rate keeping the smallest error, still requiring less computational time for the most cases.

A non-parametric statistical significance proof called Wilcoxon's rank sum test for independent samples [37–39] has been conducted at the 5% significance level on the multiple error (ME) data of Table 3.2. Table 3.3 reports the p-values produced by Wilcoxon's test for the pair-wise comparison of the multiple error (ME) of two

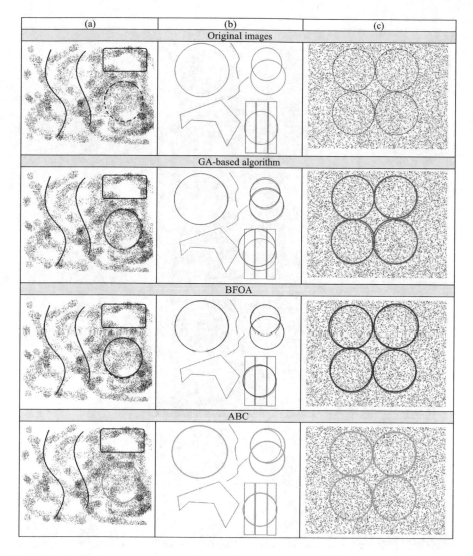

Fig. 3.11 Synthetic images and their detected circles for: GA-based algorithm, the BFOA method and the proposed ABC algorithm

groups. One group corresponding to ABC versus GA and the other corresponds to an ABC versus BFOA, one at a time. As a null hypothesis, it is assumed that there is no significant difference between the mean values of two groups. The alternative hypothesis considers a significant difference between mean values of both groups. All p-values reported in the table are less than 0.05 (5% significance level) which is a

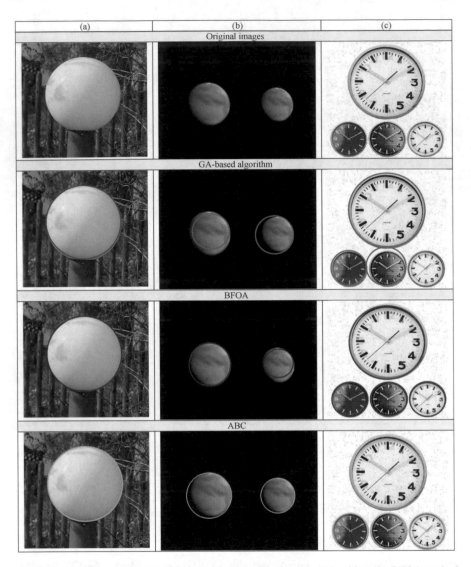

Fig. 3.12 Real-life images and their detected circles for: GA-based algorithm, the BFOA method and the proposed ABC algorithm

strong evidence against the null hypothesis, indicating that the best ABC mean values for the performance are statistically significant which has not occurred by chance.

Figure 3.13 demonstrates the relative performance of ABC in comparison to the RHT algorithm as it is described in [14]. Images from the test are complicated and also contain different noise conditions. Table 3.4 reports the corresponding

Table 3.2 The averaged execution-time, success rate and the averaged multiple error for the GA-based algorithm, the BFOA method and the proposed ABC algorithm, considering six test images (shown by Figs. 3.8 and 3.9)

Image	Averaged execution time ± standard deviation (s)			Success rate (SR) (%)			Averaged ME ± standard deviation		
	GA	BFOA	ABC	GA	BFOA	ABC	GA	BFOA	ABC
Synthetic images									
(a)	2.23 ± (0.41)	1.71 ± (0.51)	**0.21 ± (0.22)**	94	**100**	**100**	0.41 ± (0.044)	0.33 ± (0.052)	**0.22 ± (0.033)**
(b)	3.15 ± (0.39)	2.80 ± (0.65)	**0.36 ± (0.24)**	81	95	**98**	0.51 ± (0.038)	0.37 ± (0.032)	**0.26 ± (0.041)**
(c)	4.21 ± (0.11)	3.18 ± (0.36)	**0.20 ± (0.19)**	79	91	**100**	0.48 ± (0.029)	0.41 ± (0.051)	**0.15 ± (0.036)**
Natural images									
(a)	5.11 ± (0.43)	3.45 ± (0.52)	**1.10 ± (0.24)**	93	**100**	**100**	0.45 ± (0.051)	0.41 ± (0.029)	**0.25 ± (0.037)**
(b)	6.33 ± (0.34)	4.11 ± (0.14)	**1.61 ± (0.17)**	87	94	**100**	0.81 ± (0.042)	0.77 ± (0.051)	**0.37 ± (0.055)**
(c)	7.62 ± (0.97)	5.36 ± (0.17)	**1.95 ± (0.41)**	88	90	**98**	0.92 ± (0.075)	0.88 ± (0.081)	**0.41 ± (0.066)**

Table 3.3 *p*-values produced by Wilcoxon's test comparing ABC to GA and BFOA over the averaged ME from Table 3.2

Image	*p*-value	
	ABC versus GA	ABC versus BFOA
Synthetic images		
(a)	1.8061e−004	1.8288e−004
(b)	1.7454e−004	1.9011e−004
(c)	1.7981e−004	1.8922e−004
Natural images		
(a)	1.7788e−004	1.8698e−004
(b)	1.6989e−004	1.9124e−004
(c)	1.7012e−004	1.9081e−004

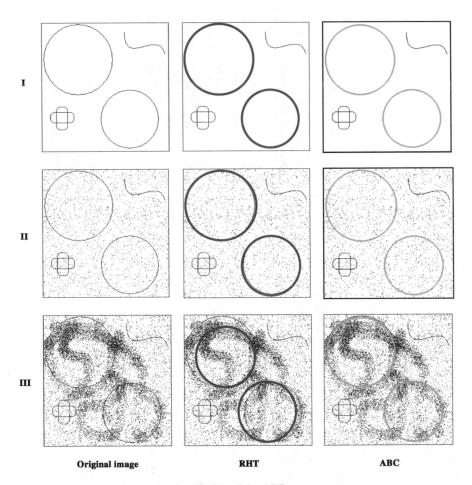

Fig. 3.13 Relative performance of the RHT and the ABC

Table 3.4 Average time, success rate and averaged error for the ABC and the HT, considering three test images

Image	Average time ± standard deviation (s)		Success rate (SR) (%)		Average ME ± standard deviation	
	RHT	ABC	RHT	ABC	RHT	ABC
(I)	7.82 ± (0.34)	**0.20 ± (0.31)**	**100**	**100**	0.19 ± (0.041)	**0.20 ± (0.021)**
(II)	8.65 ± (0.48)	**0.23 ± (0.28)**	70	**100**	0.47 ± (0.037)	**0.18 ± (0.035)**
(III)	10.65 ± (0.48)	**0.22 ± (0.21)**	28	**98**	1.21 ± (0.033)	**0.23 ± (0.028)**

averaged execution time, success rate (in %), and average multiple error (calculated following Eq. 3.16) for ABC and RHT algorithms over three test images shown by Fig. 3.10. Table 3.4 shows a performance loss as noise conditions change. Yet the ABC algorithm holds its performance under the same circumstances.

3.6 Conclusions

This chapter has presented an algorithm for the automatic detection of multiple circular shapes from complicated and noisy images with no consideration of conventional Hough transform principles. The detection process is considered to be similar to a multi-modal optimization problem. In contrast to other heuristic methods that employ an iterative procedure, the proposed ABC method is able to detect single or multiple circles over a digital image considering only one optimization cycle. The ABC algorithm searches the entire edge-map for circular shapes by using a combination of three non-collinear edge points as candidate circles (food positions) in the edge-only image. A matching function (objective function) is used to measure the existence of a candidate circle over the edge map. Guided by the values of this matching function, the set of encoded candidate circles are evolved using the ABC algorithm so that the best candidate can fit into an actual circle. A novel contribution is related to the exhausted-source memory which has been designed to hold "stuck" solutions which in turn represent feasible solutions for the multi-circle detection problem. A post-analysis on the exhausted-source memory should indeed detect other local minima, i.e. other potential circular shapes. The overall approach generates a fast sub-pixel detector which can effectively identify multiple circles in real images despite circular objects exhibiting a significant occluded portion.

Classical Hough Transform methods for circle detection use three edge points to cast a vote for the potential circular shape in the parameter space. However, they would require a huge amount of memory and longer computational times to obtain a sub-pixel resolution. Moreover, HT-based methods rarely find a precise parameter set for a circle in the image [32]. In our approach, the detected circles are directly obtained from Eqs. 3.6–3.9, still reaching sub-pixel accuracy.

In order to test the circle detection performance, speed and accuracy have been both compared. A score function to measure accuracy is defined by Eq. (3.11) which can effectively evaluate the mismatch between a manually-detected and a machine-detected circle. We have demonstrated that the ABC method outperforms both the GA (as described in [17]) and the BFOA (as described in [18]) within a statistically significant framework (Wilcoxon test). In contrast to the ABC method, the RHT algorithm [14] has shown a performance loss under noisy conditions. Yet the ABC algorithm holds its performance under the same circumstances.

Table 3.2 indicates that the ABC method can yield better results on complicated and noisy images in comparison to the GA and the BFOA methods. However, this chapter does not aim to beat all the circle detector methods proposed earlier, but to show that the ABC algorithm can effectively serve as an attractive method to successfully extract multiple circular shapes.

References

1. da Fontoura Costa, L., Marcondes Cesar Jr., R., 2001. Shape Análisis and Classification. CRC Press, Boca Raton FL.
2. Yuen, H., Princen, J., Illingworth, J., Kittler, J., 1990. Comparative study of Hough transform methods for circle finding. Image Vision Comput. 8 (1), 71–77.
3. Iivarinen, J., Peura, M., Sarela, J., Visa, A., 1997. Comparison of combined shape descriptors for irregular objects. In: Proc. 8th British Machine Vision Conf., Cochester, UK, pp. 430–439.
4. Jones, G., Princen, J., Illingworth, J., Kittler, J., 1990. Robust estimation of shape parameters. In: Proc. British Machine Vision Conf., pp. 43–48.
5. Fischer, M., Bolles, R., 1981. Random sample consensus: A paradigm to model fitting with applications to image analysis and automated cartography. CACM 24 (6), 381–395.
6. Bongiovanni, G., and Crescenzi, P.: Parallel Simulated Annealing for Shape Detection, *Computer Vision and Image Understanding*, vol. 61, no. 1, pp. 60–69, 1995.
7. Roth, G., Levine, M.D., 1994. Geometric primitive extraction using a genetic algorithm. IEEE Trans. Pattern Anal. Machine Intell. 16 (9), 901–905.
8. Cuevas, E., Zaldivar, D., Pérez-Cisneros, M., Ramírez-Ortegón, M., Circle detection using discrete differential evolution Optimization, Pattern Analysis and Applications, 14 (1), (2011), 93–107.
9. Cuevas, E., Ortega-Sánchez, N., Zaldivar, D., Pérez-Cisneros, M., Circle detection by Harmony Search Optimization, Journal of Intelligent and Robotic Systems: Theory and Applications, 66 (3), (2012), 359–376.
10. Peura, M., Iivarinen, J., 1997. Efficiency of simple shape descriptors. In: Arcelli, C., Cordella, L.P., di Baja, G.S. (Eds.), Advances in Visual Form Analysis. World Scientific, Singapore, pp. 443–451.
11. Muammar, H., Nixon, M., 1989. Approaches to extending the Hough transform. In: Proc. Int. Conf. on Acoustics, Speech and Signal Processing ICASSP_89, vol. 3, pp. 1556–1559.
12. T.J. Atherton, D.J. Kerbyson, "Using phase to represent radius in the coherent circle Hough transform", *Proc, IEE Colloquium on the Hough Transform*, IEE, London, 1993.
13. Shaked, D., Yaron, O., Kiryati, N., 1996. Deriving stopping rules for the probabilistic Hough transform by sequential analysis. Comput. Vision Image Understanding 63, 512–526.

14. Xu, L., Oja, E., Kultanen, P., 1990. A new curve detection method: Randomized Hough transform (RHT). Pattern Recognition Lett. 11 (5), 331–338.
15. Han, J.H., Koczy, L.T., Poston, T., 1993. Fuzzy Hough transform. In: Proc. 2nd Int. Conf. on Fuzzy Systems, vol. 2, pp. 803–808.
16. Becker J., Grousson S., Coltuc D., 2002. From Hough transforms to integral transforms. In: Proc. Int. Geoscience and Remote Sensing Symp., 2002 IGARSS_02, vol. 3, pp. 1444–144.
17. Ayala-Ramirez, V., Garcia-Capulin, C. H., Perez-Garcia, A. and Sanchez-Yanez, R. E. Circle detection on images using genetic algorithms. Pattern Recognition Letters, 2006, 27, pp. 652–657.
18. Dasgupta, S., Das, S., Biswas A. and Abraham, A. Automatic circle detection on digital images whit an adaptive bacterial foraging algorithm. Soft Computing, 2009, doi:10.1007/s00500-009-0508-z.
19. Holland, J.H., Adaptation in Natural and Artificial Systems, University of Michigan Press, Ann Arbor, MI, 1975.
20. K. Price, R. Storn, A. Lampinen, Differential Evolution a Practical Approach to Global Optimization, Springer Natural Computing Series, 2005.
21. J. Kennedy, R. Eberhart, Particle swarm optimization, in: IEEE International Conference on Neural Networks (Piscataway, NJ), 1995, pp. 1942 1948.
22. M. Dorigo, V. Maniezzo, A. Colorni, Positive feedback as a search strategy, Technical Report 91-016, Politecnico di Milano, Italy, 1991.
23. Liu Y, Passino K. Biomimicry of social foraging bacteria for distributed optimization: models, principles, and emergent behaviors. J Optim Theory Appl 115(3):603–628, 2002.
24. D. Karaboga. An idea based on honey bee swarm for numerical optimization, technical report-TR06, Erciyes University, Engineering Faculty, Computer Engineering Department 2005.
25. N. Karaboga. A new design method based on artificial bee colony algorithm for digital IIR filters. Journal of the Franklin Institute 346 (2009) 328–348.
26. S. L. Ho, S. Yang. An artificial bee colony algorithm for inverse problems. International Journal of Applied Electromagnetics and Mechanics, 31 (2009) 181–192.
27. Ying-ping Chen, Pei Jiang. Analysis of particle interaction in particle swarm optimization. Theoretical Computer Science 411(21), 2010, 2101–2115.
28. Hongnian Zang, Shujun Zhang, Kevin Hapeshia. A Review of Nature-Inspired Algorithms. Journal of Bionic Engineering 7(1), 2010, S232–S237.
29. Josef Tvrdík. Adaptation in differential evolution: A numerical comparison. Applied Soft Computing 9(3), 2009, 1149–1155.
30. Arijit Biswas, Swagatam Das, Ajith Abraham, Sambarta Dasgupta. Stability analysis of the reproduction operator in bacterial foraging optimization. Theoretical Computer Science 411, 2010, 2127–2139.
31. D. Karaboga, B. Basturk. On the performance of artificial bee colony (ABC) algorithm. Applied soft computing, Volume 8, Issue 1, January 2008, Pages 687–697.
32. D. Karaboga, B. Akay. A comparative study of Artificial Bee Colony algorithm. Applied Mathematics and Computation 214 (2009) 108–132.
33. Bresenham, J.E.: A Linear Algorithm for Incremental Digital Display of Circular Arcs. Communications of the ACM 20, 100–106. (1987).
34. Van Aken, J R. Efficient ellipse-drawing algorithm, IEEE Comp, Graphics applic., 2005. 4, (9), pp. 24–35.
35. Aytug, H., Koehler, G.J.: New stopping criterion for genetic algorithms. European Journal of Operational Research 126 (2000) 662–674.
36. Greenhalgh, D., Marshall, S.: Convergence criteria for genetic algorithms. SIAM Journal on Computing 20 (2000) 269–282.
37. Wilcoxon F (1945) Individual comparisons by ranking methods. Biometrics 1:80–83.

38. Garcia S, Molina D, Lozano M, Herrera F (2008) A study on the use of non-parametric tests for analyzing the evolutionary algorithms' behaviour: a case study on the CEC'2005 Special session on real parameter optimization. J Heurist. doi:10.1007/s10732-008-9080-4.
39. J. Santamaría, O. Cordón, S. Damas, J.M. García-Torres, A. Quirin, Performance Evaluation of Memetic Approaches in 3D Reconstruction of Forensic Objects. Soft Computing, doi:10.1007/s00500-008-0351-7, in press (2008).

Chapter 4
Template Matching

Abstract Template matching (TM) plays an important role in several image processing applications. In a TM approach, it is sought the point in which it is presented the best possible resemblance between a sub-image known as template and its coincident region within a source image. TM involves two critical aspects: similarity measurement and search strategy. The simplest available TM method finds the best possible coincidence between the images through an exhaustive computation of the Normalized cross-correlation (NCC) values (similarity measurement) for all elements of the source image (search strategy). Recently, several TM algorithms, based on evolutionary approaches, have been proposed to reduce the number of NCC operations by calculating only a subset of search locations. On the other hand, bio-inspired computing has demonstrated to be useful in several application areas. Over the last decade, new bio-inspired algorithms have emerged with applications for detection, optimization and classification for its use in image processing. In this chapter, the Social Spider Optimization (SSO) algorithm is presented to reduce the number of search locations in the TM process. The SSO algorithm is based on the simulation of cooperative behavior of social-spiders. The algorithm considers two different search individuals (spiders): males and females. Depending on gender, each individual is conducted by a set of different evolutionary operators which mimic different cooperative behaviors that are typically found in the colony. In the proposed approach, spiders represent search locations which move throughout the positions of the source image. The NCC coefficient, used as a fitness value, evaluates the matching quality presented between the template image and the coincident region of the source image, for a determined search position (spider). The number of NCC evaluations is reduced by considering a memory which stores the NCC values previously visited in order to avoid the re-evaluation of the same search locations. Guided by the fitness values (NCC coefficients), the set of encoded candidate positions are evolved using the SSO

© Springer International Publishing AG 2017
E. Cuevas et al., *Evolutionary Computation Techniques:*
A Comparative Perspective, Studies in Computational Intelligence 686,
DOI 10.1007/978-3-319-51109-2_4

65

operators until the best possible resemblance has been found. Conducted simulations show that the proposed method achieves the best balance over other TM algorithms, in terms of estimation accuracy and computational cost.

4.1 Introduction

Bio-inspired computing [1] is concerned with the use of biology as an inspiration for solving computational problems. The increasing interest in this field lies in the fact that nowadays the world is facing more and more complex, large, distributed and ill-structured systems, while on the other hand, people notice that the apparently simple structures and organizations in nature are capable of dealing with most complex systems and tasks with ease.

The interesting and exotic collective behavior of social insects have fascinated and attracted researchers for many years. The collaborative swarming behavior observed in these groups provides survival advantages, where insect aggregations of relatively simple and "unintelligent" individuals can accomplish very complex tasks using only limited local information and simple rules of behavior [2]. Social-spiders are a representative example of social insects [3]. A social-spider is a spider species whose members maintain a set of complex cooperative behaviors [4]. Whereas most spiders are solitary and even aggressive toward other members of their own species, social-spiders show a tendency to live in groups, forming long-lasting aggregations often referred to as colonies [5]. In a social-spider colony, each member, depending on its gender, executes a variety of tasks such as predation, mating, web design, and social interaction [5, 6]. The web it is an important part of the colony because it is not only used as a common environment for all members, but also as a communication channel among them [7]. Therefore, important information (such as trapped prays or mating possibilities) is transmitted by small vibrations through the web. Such information, considered as a local knowledge, is employed by each member to conduct its own cooperative behavior, influencing simultaneously the social regulation of the colony [8].

On the other hand, Template matching (TM) is an image processing technique to find object in images. In a TM approach, it is sought the point in which it is presented the best possible resemblance between a sub-image known as template and its coincident region within a source image.

In general TM involves two critical points: the similarity measurement and the search strategy [9]. Several metrics have been proposed to evaluate the matching between two images, the most important are: sum of absolute differences (SAD), sum of squared differences (SSD) and the normalized cross-correlation (NCC). The most used matching criterion is the NCC coefficient which is computationally expensive and represents the most consuming operation in the TM process [10].

The full search algorithm is the simplest TM algorithm that can deliver the optimal detection with respect to a maximal NCC coefficient as it checks all pixel-candidates one at a time. Unfortunately, such exhaustive search and the NCC calculation at each checking point, yields an extremely computational expensive TM method that seriously constraints its use for several image processing applications.

Recently, several TM algorithms, based on evolutionary approaches, have been proposed to reduce the number of NCC operations by calculating only a subset of search locations. Such approaches have produced several robust detectors using different optimization methods such as Genetic algorithms (GA) [11], Particle Swarm Optimization (PSO) [12, 13] and Imperialist competitive algorithm (ICA) [14]. Although these algorithms allow reducing the number of search locations, they do not explore the whole region effectively and often suffers premature convergence which conducts to sub-optimal detections. The reason of these problems is the operators used for modifying the particles. In such algorithms, during their evolution, the position of each agent in the next iteration is updated yielding an attraction towards the position of the best particle seen so-far [15, 16]. This behavior produces that the entire population, as the algorithm evolves, concentrates around the best particle, favoring the premature convergence and damaging the particle diversity.

One particular difficulty in applying an evolutionary algorithm to discrete optimization problems, such TM, is the multiple evaluation of the same individual. Discrete optimization problems are defined by using search spaces compound by a set of finite solutions. Therefore, since random numbers are involved in the calculation of new individuals, they may encounter the same solutions (repetition) that have been visited by other individuals at previous iterations, particularly when individuals are confined to a finite area. Evidently, such fact seriously constraints its performance mainly when fitness evaluation is computationally expensive to calculate.

In this chapter, the novel bio-inspired algorithm called the Social Spider Optimization (SSO) is presented to reduce the number of search locations in the TM process. The SSO algorithm is based on the simulation of cooperative behavior of social-spiders. The algorithm considers two different search individuals (spiders): males and females. Depending on gender, each individual is conducted by a set of different evolutionary operators which mimic different cooperative behaviors that are typically found in the colony. In the proposed approach, spiders represent search locations which move throughout the positions of the source image. The NCC coefficient, used as a fitness value, evaluates the matching quality presented between the template image and the coincident region of the source image, for a determined search position (spider). The number of NCC evaluations is reduced by considering a memory which stores the NCC values previously visited in order to avoid the re-evaluation of the same search locations. Guided by the fitness values (NCC coefficients), the set of encoded candidate positions are evolved using the SSO operators until the best possible resemblance has been found. To illustrate the proficiency and robustness of the proposed algorithm, it is compared to other recently published evolutionary methods, employed to solve the TM problem. The comparison examines several standard benchmark images. Experimental results show that the proposed method achieves the best balance, in terms of both estimation accuracy and computational cost.

The overall chapter is organized as follows: In Sect. 4.2, we introduce basic biological aspects of the algorithm. Section 4.3 holds a description about the SSO algorithm. Section 4.4 provides backgrounds about the TM process while Sect. 4.5 exposes the final TM algorithm as a combination of SSO and the NCC evaluation. Section 4.6 demonstrates experimental results for the proposed approach over standard test images and some conclusions are drawn in Sect. 4.7.

4.2 Biological Fundamentals

Social insect societies are complex cooperative systems that self-organize within a set of constraints. Cooperative groups are better at manipulating and exploiting their environment, defending resources and brood, and allowing task specialization among group members [17, 18]. A social insect colony functions as an integrated unit that not only possesses the ability to operate at a distributed manner, but also to undertake enormous construction of global projects [19]. It is important to acknowledge that global order in social insects can arise as a result of internal interactions among members.

A few species of spiders have been documented exhibiting a degree of social behavior [3]. The behavior of spiders can be generalized into two basic forms: solitary spiders and social spiders [5]. This classification is made based on the level of cooperative behavior that they exhibit [6]. In one side, solitary spiders create and maintain their own web while live in scarce contact to other individuals of the same species. In contrast, social spiders form colonies that remain together over a communal web with close spatial relationship to other group members [7].

A social spider colony is composed of two fundamental components: its members and the communal web. Members are divided into two different categories: males and females. An interesting characteristic of social-spiders is the highly female-biased population. Some studies suggest that the number of male spiders barely reaches the 30% of the total colony members [5, 20]. In the colony, each member, depending on its gender, cooperate in different activities such as building and maintaining the communal web, prey capturing, mating and social contact [20]. Interactions among members are either direct or indirect [21]. Direct interactions imply body contact or the exchange of fluids such as mating. For indirect interactions, the communal web is used as a "medium of communication" which conveys important information that is available to each colony member [7]. This information encoded as small vibrations is a critical aspect for the collective coordination among members [8]. Vibrations are employed by the colony members to decode several messages such as the size of the trapped preys, characteristics of the neighboring members, etc. The intensity of such vibrations depend on the weight and distance of the spiders that have produced them.

In spite of the complexity, all the cooperative global patterns in the colony level are generated as a result of internal interactions among colony members [22]. Such internal interactions involve a set of simple behavioral rules followed by each spider in the colony. Behavioral rules are divided into two different classes: social inter-action (cooperative behavior) and mating [23].

As a social insect, spiders perform cooperative interaction with other colony members. The way in which this behavior takes place depends on the spider gender. Female spiders which show a major tendency to socialize present an attraction or dislike over others, irrespectively of gender [5]. For a particular female spider, such attraction or dislike is commonly developed over other spiders according to their vibrations which are emitted over the communal web and represent strong colony members [8]. Since the vibrations depend on the weight and distance of the members which provoke them, stronger vibrations are produced either by big spiders or neighboring members [7]. The bigger a spider is, the better it is considered as a colony member. The final decision of attraction or dislike over a determined member is taken according to an internal state which is influenced by several factors such as reproduction cycle, curiosity and other random phenomena [8].

Different to female spiders, the behavior of male members is reproductive-oriented [24]. Male spiders recognize themselves as a subgroup of alpha males which dominate the colony resources. Therefore, the male population is divided into two classes: dominant and non-dominant male spiders [24]. Dominant male spiders have better fitness characteristics (normally size) in comparison to non-dominant. In a typical behavior, dominant males are attracted to the closest female spider in the communal web. In contrast, non-dominant male spiders tend to concentrate upon the center of the male population as a strategy to take advantage of the resources wasted by dominant males [25].

Mating is an important operation that no only assures the colony survival, but also allows the information exchange among members. Mating in a social-spider colony is performed by dominant males and female members [26]. Under such circum-stances, when a dominant male spider locates one or more female members within a specific range, it mates with all the females in order to produce offspring [27].

4.3   The Social Spider Optimization (SSO) Algorithm

In this chapter, the operational principles from the social-spider colony have been used as guidelines for developing a new swarm optimization algorithm. The SSO assumes that entire search space is a communal web, where all the social-spiders interact to each other. In the proposed approach, each solution within the search space represents a spider position in the communal web. Every spider receives a

weight according to the fitness value of the solution that is symbolized by the
social-spider. The algorithm models two different search agents (spiders): males and
females. Depending on gender, each individual is conducted by a set of different
evolutionary operators which mimic different cooperative behaviors that are commonly assumed within the colony.

An interesting characteristic of social-spiders is the highly female-biased populations. In order to emulate this fact, the algorithm starts by defining the number of
female and male spiders that will be characterized as individuals in the search
space. The number of females N_f is randomly selected within the range of 65–90%
of the entire population N. Therefore, N_f is calculated by the following equation:

$$N_f = \text{floor}[(0.9 - \text{rand} \cdot 0.25) \cdot N] \tag{4.1}$$

where rand is a random number between [0,1] whereas floor($\cdot$) maps a real number
to an integer number. The number of male spiders N_m is computed as the complement between N and N_f. It is calculated as follows:

$$N_m = N - N_f \tag{4.2}$$

Therefore, the complete population $\mathbf{S}$, composed by N elements, is divided in
two sub-groups $\mathbf{F}$ and $\mathbf{M}$. The Group $\mathbf{F}$ assembles the set of female individuals
($\mathbf{F} = \{\mathbf{f}_1, \mathbf{f}_2, \ldots, \mathbf{f}_{N_f}\}$) whereas $\mathbf{M}$ groups the male members ($\mathbf{M} = \{\mathbf{m}_1, \mathbf{m}_2, \ldots,$
$\mathbf{m}_{N_m}\}$),   where $\mathbf{S} = \mathbf{F} \cup \mathbf{M}$ ($\mathbf{S} = \{\mathbf{s}_1, \mathbf{s}_2, \ldots, \mathbf{s}_N\}$),   such   that $\mathbf{S} = \{\mathbf{s}_1 = \mathbf{f}_1,$
$\mathbf{s}_2 = \mathbf{f}_2, \ldots, \mathbf{s}_{N_f} = \mathbf{f}_{N_f}, \mathbf{s}_{N_f+1} = \mathbf{m}_1, \mathbf{s}_{N_f+2} = \mathbf{m}_2, \ldots, \mathbf{s}_N = \mathbf{m}_{N_m}\}$.

4.3.1 Fitness Assignation

In the biological metaphor, the spider size is the characteristic that evaluates the
individual capacity to perform better over its assigned tasks. In the proposed approach,
every individual (spider) receives a weight w_i which represents the solution quality
that corresponds to the spider i (irrespective of gender) of the population $\mathbf{S}$. In order to
calculate the weight of every spider the next equation is used:

$$w_i = \frac{J(\mathbf{s}_i) - worst_{\mathbf{S}}}{best_{\mathbf{S}} - worst_{\mathbf{S}}} \tag{4.3}$$

where $J(\mathbf{s}_i)$ is the fitness value obtained by the evaluation of the spider position $\mathbf{s}_i$
with regard to the objective function $J(\cdot)$. The values $worst_{\mathbf{S}}$ and $best_{\mathbf{S}}$ are defined
as follows (considering a maximization problem):

$$best_{\mathbf{S}} = \max_{k \in \{1,2,\ldots,N\}} (J(\mathbf{s}_k)) \quad \text{and} \quad worst_{\mathbf{S}} = \min_{k \in \{1,2,\ldots,N\}} (J(\mathbf{s}_k)) \tag{4.4}$$

4.3.2 Modeling of the Vibrations Through the Communal Web

The communal web is used as a mechanism to transmit information among the colony members. This information is encoded as small vibrations that are critical for the collective coordination of all individuals in the population. The vibrations depend on the weight and distance of the spider which has generated them. Since the distance is relative to the individual that provokes the vibrations and the member who detects them, members located near to the individual that provokes the vibrations, perceive stronger vibrations in comparison with members located in distant positions. In order to reproduce this process, the vibrations perceived by the individual i as a result of the information transmitted by the member j are modeled according to the following equation:

$$Vib_{i,j} = w_j \cdot e^{-d_{i,j}^2} \qquad (4.5)$$

where the $d_{i,j}$ is the Euclidian distance between the spiders i and j, such that $d_{i,j} = \|\mathbf{s}_i - \mathbf{s}_j\|$.

Although it is virtually possible to compute perceived-vibrations by considering any pair of individuals, three special relationships are considered within the SSO approach:

1. Vibrations $Vibc_i$ are perceived by the individual i ($\mathbf{s}_i$) as a result of the information transmitted by the member c ($\mathbf{s}_c$) who is an individual that has two important characteristics: it is the nearest member to i and possesses a higher weight in comparison to i ($w_c > w_i$).

$$Vibc_i = w_c \cdot e^{-d_{i,c}^2} \qquad (4.6)$$

2. The vibrations $Vibb_i$ perceived by the individual i as a result of the information transmitted by the member b ($\mathbf{s}_b$), with b being the individual holding the best weight (best fitness value) of the entire population $\mathbf{S}$, such that $w_b = \max_{k \in \{1,2,\ldots,N\}} (w_k)$.

$$Vibb_i = w_b \cdot e^{-d_{i,b}^2} \qquad (4.7)$$

3. The vibrations $Vibf_i$ perceived by the individual i ($\mathbf{s}_i$) as a result of the information transmitted by the member f ($\mathbf{s}_f$), with f being the nearest female individual to i.

$$Vibf_i = w_f \cdot e^{-d_{i,f}^2} \qquad (4.8)$$

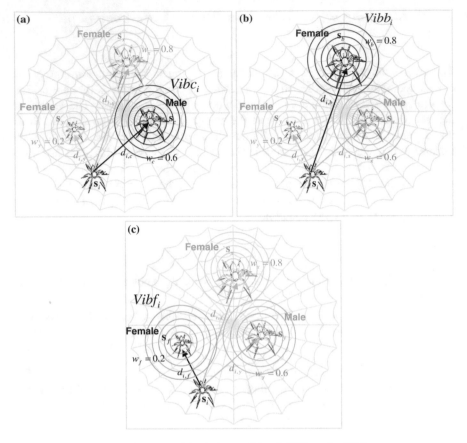

Fig. 4.1 Configuration of each special relation: **a** $Vibc_i$, **b** $Vibb_i$ and **c** $Vibf_i$

Figure 4.1 shows the configuration of each special relationship: (a) $Vibc_i$, (b) $Vibb_i$ and (c) $Vibf_i$.

4.3.3   Initializing the Population

Like other evolutionary algorithms, the SSO is an iterative process whose first step is to randomly initialize the entire population (female and male). The algorithm begins by initializing the set **S** of N spider positions. Each spider position, $\mathbf{f}_i$ or $\mathbf{m}_i$, is a n-dimensional vector containing the parameter values to be optimized. Such values are randomly and uniformly distributed between the pre-specified lower initial parameter bound p_j^{low} and the upper initial parameter bound p_j^{high}, just as it described by the following expressions:

$$f_{i,j}^0 = p_j^{low} + \text{rand}(0,1) \cdot (p_j^{high} - p_j^{low}) \quad i = 1,2,\ldots,N_f; \quad j = 1,2,\ldots,n$$
$$m_{k,j}^0 = p_j^{low} + \text{rand}(0,1) \cdot (p_j^{high} - p_j^{low}) \quad k = 1,2,\ldots,N_m; \quad j = 1,2,\ldots,n \tag{4.9}$$

where j, i and k are the parameter and individual indexes respectively whereas zero signals the initial population. The function rand(0,1) generates a random number between 0 and 1. Hence, $f_{i,j}$ is the jth parameter of the ith female spider position.

4.3.4   Cooperative Operators

4.3.4.1   Female Cooperative Operator

Social-spiders perform cooperative interaction over other colony members. The way in which this behavior takes place depends on the spider gender. Female spiders present an attraction or dislike over others irrespective of gender. For a particular female spider, such attraction or dislike is commonly developed over other spiders according to their vibrations which are emitted over the communal web. Since vibrations depend on the weight and distance of the members which have originated them, strong vibrations are produced either by big spiders or other neighboring members lying nearby the individual which is perceiving them. The final decision of attraction or dislike over a determined member is taken considering an internal state which is influenced by several factors such as reproduction cycle, curiosity and other random phenomena.

In order to emulate the cooperative behavior of the female spider, a new operator is defined. The operator considers the position change of the female spider i at each iteration. Such position change, which can be of attraction or repulsion, is computed as a combination of three different elements. The first one involves the change in regard to the nearest member to i that holds a higher weight and produces the vibration $Vibc_i$. The second one considers the change regarding the best individual of the entire population S who produces the vibration $Vibb_i$. Finally, the third one incorporates a random movement.

Since the final movement of attraction or repulsion depends on several random phenomena, the selection is modeled as a stochastic decision. For this operation, a uniform random number r_m is generated within the range [0,1]. If r_m is smaller than a threshold PF, an attraction movement is generated; otherwise, a repulsion movement is produced. Therefore, such operator can be modeled as follows:

$$\mathbf{f}_i^{k+1} = \begin{cases} \mathbf{f}_i^k + \alpha \cdot Vibc_i \cdot (\mathbf{s}_c - \mathbf{f}_i^k) + \beta \cdot Vibb_i \cdot (\mathbf{s}_b - \mathbf{f}_i^k) + \delta \cdot (\text{rand} - \frac{1}{2}) & \text{with probability } PF \\ \mathbf{f}_i^k - \alpha \cdot Vibc_i \cdot (\mathbf{s}_c - \mathbf{f}_i^k) - \beta \cdot Vibb_i \cdot (\mathbf{s}_b - \mathbf{f}_i^k) + \delta \cdot (\text{rand} - \frac{1}{2}) & \text{with probability } 1 - PF \end{cases}$$

$$(4.10)$$

where α, β, δ and rand are random numbers between [0,1] whereas k represents the iteration number. The individual $\mathbf{s}_c$ and $\mathbf{s}_b$ represent the nearest member to i that holds a higher weight and the best individual of the entire population $\mathbf{S}$, respectively.

Under this operation, each particle presents a movement which combines the past position that holds the attraction or repulsion vector over the local best element $\mathbf{s}_c$ and the global best individual $\mathbf{s}_b$ seen so-far. This particular type of interaction avoids the quick concentration of particles at only one point and encourages each particle to search around the local candidate region within its neighborhood ($\mathbf{s}_c$), rather than interacting to a particle ($\mathbf{s}_b$) in a distant region of the domain. The use of this scheme has two advantages. First, it prevents the particles from moving towards the global best position, making the algorithm less susceptible to premature convergence. Second, it encourages particles to explore their own neighborhood thoroughly before converging towards the global best position. Therefore, it provides the algorithm with global search ability and enhances the exploitative behavior of the proposed approach.

4.3.4.2 Male Cooperative Operator

According to the biological behavior of the social-spider, male population is divided into two classes: dominant and non-dominant male spiders. Dominant male spiders have better fitness characteristics (usually regarding the size) in comparison to non-dominant. Dominant males are attracted to the closest female spider in the communal web. In contrast, non-dominant male spiders tend to concentrate in the center of the male population as a strategy to take advantage of resources that are wasted by dominant males.

For emulating such cooperative behavior, the male members are divided into two different groups (dominant members $\mathbf{D}$ and non-dominant members $\mathbf{ND}$) according to their position with regard to the median member. Male members, with a weight value above the median value within the male population, are considered the dominant individuals $\mathbf{D}$. On the other hand, those under the median value are labeled as non-dominant $\mathbf{ND}$ males. In order to implement such computation, the male population $\mathbf{M}$ ($\mathbf{M} = \{\mathbf{m}_1, \mathbf{m}_2, \ldots, \mathbf{m}_{N_m}\}$) is arranged according to their weight value in decreasing order. Thus, the individual whose weight $w_{N_f + m}$ is located in the middle is considered the median male member. Since indexes of the male population $\mathbf{M}$ in regard to the entire population $\mathbf{S}$ are increased by the number of female members N_f, the median weight is indexed by $N_f + m$. According to this, change of positions for the male spider can be modeled as follows:

$$\mathbf{m}_i^{k+1} = \begin{cases} \mathbf{m}_i^k + \alpha \cdot Vibf_i \cdot (\mathbf{s}_f - \mathbf{m}_i^k) + \delta \cdot (\text{rand} - \frac{1}{2}) & \text{if } w_{N_f+i} > w_{N_f+m} \\ \mathbf{m}_i^k + \alpha \cdot \left(\frac{\sum_{h=1}^{N_m} \mathbf{m}_h^k \cdot w_{N_f+h}}{\sum_{h=1}^{N_m} w_{N_f+h}} - \mathbf{m}_i^k \right) & \text{if } w_{N_f+i} \leq w_{N_f+m} \end{cases},$$

$$(4.11)$$

where the individual $\mathbf{s}_f$ represents the nearest female individual to the male member i whereas $\left(\sum_{h=1}^{N_m} \mathbf{m}_h^k \cdot w_{N_f+h} / \sum_{h=1}^{N_m} w_{N_f+h} \right)$ correspond to the weighted mean of the male population $\mathbf{M}$.

By using this operator, two different behaviors are produced. First, the set $\mathbf{D}$ of particles is attracted to others in order to provoke mating. Such behavior allows incorporating diversity into the population. Second, the set $\mathbf{ND}$ of particles is attracted to the weighted mean of the male population $\mathbf{M}$. This fact is used to partially control the search process according to the average performance of a sub-group of the population. Such mechanism acts as a filter which avoids that very good individuals or extremely bad individuals influence the search process.

4.3.5 Mating Operator

Mating in a social-spider colony is performed by dominant males and the female members. Under such circumstances, when a dominant male $\mathbf{m}_g$ spider ($g \in \mathbf{D}$) locates a set $\mathbf{E}^g$ of female members within a specific range r (range of mating), it mates, forming a new brood $\mathbf{s}_{new}$ which is generated considering all the elements of the set $\mathbf{T}^g$ that, in turn, has been generated by the union $\mathbf{E}^g \cup \mathbf{m}_g$. It is important to emphasize that if the set $\mathbf{E}^g$ is empty, the mating operation is canceled. The range r is defined as a radius which depends on the size of the search space. Such radius r is computed according to the following model:

$$r = \frac{\sum_{j=1}^{n} (p_j^{high} - p_j^{low})}{2 \cdot n} \qquad (4.12)$$

In the mating process, the weight of each involved spider (elements of $\mathbf{T}^g$) defines the probability of influence for each individual into the new brood. The spiders holding a heavier weight are more likely to influence the new product, while elements with lighter weight have a lower probability. The influence probability Ps_i of each member is assigned by the roulette method, which is defined as follows:

$$Ps_i = \frac{w_i}{\sum_{j \in \mathbf{T}^k} w_j}, \qquad (4.13)$$

where $i \in \mathbf{T}^g$.

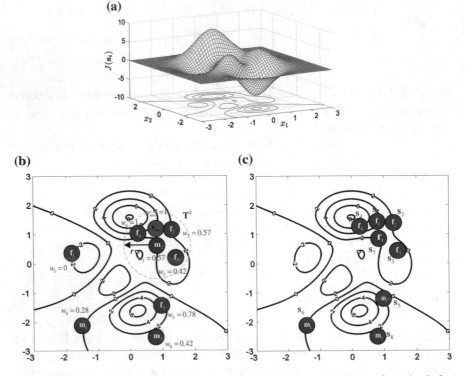

Fig. 4.2 Example of the mating operation: **a** optimization problem, **b** initial configuration before mating and **c** configuration after the mating operation

Once the new spider is formed, it is compared to the new spider candidate s_{new} holding the worst spider s_{wo} of the colony, according to their weight values (where $w_{wo} = \min_{l \in \{1,2,\ldots,N\}} (w_l)$). If the new spider is better than the worst spider, the worst spider is replaced by the new one. Otherwise, the new spider is discarded and the population does not suffer changes. In case of replacement, the new spider assumes the gender and index from the replaced spider. Such fact assures that the entire population $\mathbf{S}$ maintains the original rate between female and male members.

In order to demonstrate the mating operation, Fig. 4.2a illustrates a simple optimization problem. As an example, it is assumed a population $\mathbf{S}$ of eight different 2-dimensional members ($N = 8$), five females ($N_f = 5$) and three males ($N_m = 3$). Figure 4.2b shows the initial configuration of the proposed example with three different female members $\mathbf{f}_2(\mathbf{s}_2), \mathbf{f}_3(\mathbf{s}_3)$ and $\mathbf{f}_4(\mathbf{s}_4)$ constituting the set $\mathbf{E}^2$ which is located inside of the influence range r of a dominant male $\mathbf{m}_2(\mathbf{s}_7)$. Then, the new candidate spider s_{new} is generated from the elements $\mathbf{f}_2, \mathbf{f}_3, \mathbf{f}_4$ and $\mathbf{m}_2$ which

Table 4.1 Data for constructing the new spider s_{new} through the roulette method

Spider		Position	w_i	Ps_i	Roulette
s_1	$\mathbf{f}_1$	(-1.9,0.3)	0.00	-	
s_2	$\mathbf{f}_2$	(1.4,1.1)	0.57	0.22	
s_3	$\mathbf{f}_3$	(1.5,0.2)	0.42	0.16	
s_4	$\mathbf{f}_4$	(0.4,1.0)	1.00	0.39	
s_5	$\mathbf{f}_5$	(1.0,-1.5)	0.78	-	
s_6	$\mathbf{m}_1$	(-1.3,-1.9)	0.28	-	
s_7	$\mathbf{m}_2$	(0.9,0.7)	0.57	0.22	
s_8	$\mathbf{m}_3$	(0.8,-2.6)	0.42	-	
s_{new}		(0.9,1.1)	1.00	-	

constitute the set $\mathbf{T}^2$. Therefore, the value of the first decision variable $s_{new,1}$ for the new spider is chosen by means of the roulette mechanism considering the values already existing from the set $\{f_{2,1}, f_{3,1}, f_{4,1}, m_{2,1}\}$. The value of the second decision variable $s_{new,2}$ is also chosen in the same manner. Table 4.1 shows the data for constructing the new spider through the roulette method. Once the new spider s_{new} is formed, its weight w_{new} is calculated. As s_{new} is better than the worst member $\mathbf{f}_1$ that is present in the population $\mathbf{S}$, $\mathbf{f}_1$ is replaced by s_{new}. Therefore, s_{new} assumes the same gender and index from $\mathbf{f}_1$. Figure 4.2c shows the configuration of $\mathbf{S}$ after the mating process. Under this operation, new generated particles locally exploit the search space inside the mating range in order to find better individuals.

4.3.6   Computational Procedure

The computational procedure for the proposed algorithm can be summarized as follows:

Step 1:	Considering N as the total number of n-dimensional colony members, define the number of male N_m and females N_f spiders in the entire population $\mathbf{S}$.

$$N_f = \text{floor}\left[(0.9 - \text{rand} \cdot 0.25) \cdot N\right] \text{ and } N_m = N - N_f,$$

where rand is a random number between $[0,1]$ whereas floor$(\cdot)$ maps a real number to an integer number.

Step 2:	Initialize randomly the female ($\mathbf{F} = \{\mathbf{f}_1, \mathbf{f}_2, ..., \mathbf{f}_{N_f}\}$) and male ($\mathbf{M} = \{\mathbf{m}_1, \mathbf{m}_2, ..., \mathbf{m}_{N_m}\}$) members (where $\mathbf{S} = \left\{\mathbf{s}_1 = \mathbf{f}_1, \mathbf{s}_2 = \mathbf{f}_2, ..., \mathbf{s}_{N_f} = \mathbf{f}_{N_f}, \mathbf{s}_{N_f+1} = \mathbf{m}_1, \mathbf{s}_{N_f+2} = \mathbf{m}_2, ..., \mathbf{s}_N = \mathbf{m}_{N_m}\right\}$ and calculate the radius of mating.

$$r = \frac{\sum_{j=1}^{n}(p_j^{high} - p_j^{low})}{2 \cdot n}$$

for $(i=1; i < N_f +1; i++)$

for$(j=1; j<n+1; j++)$

$$f_{i,j}^0 = p_j^{low} + \text{rand}(0,1) \cdot (p_j^{high} - p_j^{low})$$

end for

end for

for $(k=1; k < N_m +1; k++)$

for$(j=1; j<n+1; j++)$

$$m_{k,j}^0 = p_j^{low} + \text{rand} \cdot (p_j^{high} - p_j^{low})$$

end for

end for

Step 3:	Calculate the weight of every spider of $\mathbf{S}$ (section 3.1.1).

for $(i=1, i<N+1; i++)$

$$w_i = \frac{J(\mathbf{s}_i) - worst_\mathbf{S}}{best_\mathbf{S} - worst_\mathbf{S}}$$

where $best_\mathbf{S} = \max_{k \in \{1,2,...,N\}} (J(\mathbf{s}_k))$ and $worst_\mathbf{S} = \min_{k \in \{1,2,...,N\}} (J(\mathbf{s}_k))$

end for

Step 4:	Move female spiders according to the female cooperative operator (section 3.1.4).

for $(i=1; i < N_f +1; i++)$

Calculate $Vibc_i$ and $Vibb_i$ (Section 3.1.2)

If ($r_m < PF$); where $r_m \in \text{rand}(0,1)$

$$\mathbf{f}_i^{k+1} = \mathbf{f}_i^k + \alpha \cdot Vibc_i \cdot (\mathbf{s}_c - \mathbf{f}_i^k) + \beta \cdot Vibb_i \cdot (\mathbf{s}_b - \mathbf{f}_i^k) + \delta \cdot (\text{rand} - \frac{1}{2})$$

else if

$$\mathbf{f}_i^{k+1} = \mathbf{f}_i^k - \alpha \cdot Vibc_i \cdot (\mathbf{s}_c - \mathbf{f}_i^k) - \beta \cdot Vibb_i \cdot (\mathbf{s}_b - \mathbf{f}_i^k) + \delta \cdot (\text{rand} - \frac{1}{2})$$

end if

end for

Step 5:	Move the male spiders according to the male cooperative operator (section 3.1.4).

Find the median male individual (w_{N_f+m}) from $\mathbf{M}$.

for $(i=1; i < N_m +1; i++)$

Calculate $Vibf_i$ (section 3.1.2)

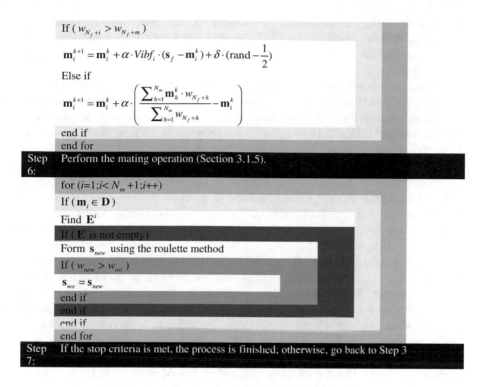

If ($w_{N_f + i} > w_{N_f + m}$)

$$\mathbf{m}_i^{k+1} = \mathbf{m}_i^k + \alpha \cdot Vibf_i \cdot (\mathbf{s}_f - \mathbf{m}_i^k) + \delta \cdot (\text{rand} - \frac{1}{2})$$

Else if

$$\mathbf{m}_i^{k+1} = \mathbf{m}_i^k + \alpha \cdot \left(\frac{\sum_{h=1}^{N_m} \mathbf{m}_h^k \cdot w_{N_f + h}}{\sum_{h=1}^{N_m} w_{N_f + h}} - \mathbf{m}_i^k \right)$$

end if
end for

Step 6: Perform the mating operation (Section 3.1.5).

for ($i=1; i< N_m + 1; i++$)
If ($\mathbf{m}_i \in \mathbf{D}$)
Find $\mathbf{E}^i$
If ($\mathbf{E}^i$ is not empty)
Form $\mathbf{s}_{new}$ using the roulette method
If ($w_{new} > w_{wo}$)
$\mathbf{s}_{wo} = \mathbf{s}_{new}$
end if
end if
end if
end for

Step 7: If the stop criteria is met, the process is finished; otherwise, go back to Step 3

4.3.7   Discussion About the SSO Algorithm

Evolutionary algorithms (EA) have been widely employed for solving complex optimization problems. These methods are found to be more powerful than conventional methods based on formal logics or mathematical programming [28]. In an EA algorithm, search agents have to decide whether to explore unknown search positions or to exploit already tested positions in order to improve their solution quality. Pure exploration degrades the precision of the evolutionary process but increases its capacity to find new potential solutions. On the other hand, pure exploitation allows refining existent solutions but adversely drives the process to local optimal solutions. Therefore, the ability of an EA to find a global optimal solutions depends on its capacity to find a good balance between the exploitation of found-so-far elements and the exploration of the search space [29]. So far, the exploration–exploitation dilemma has been an unsolved issue within the framework of evolutionary algorithms.

EA defines individuals with the same property, performing virtually the same behavior. Under these circumstances, algorithms waste the possibility to add new and selective operators as a result of considering individuals with different characteristics. These operators could incorporate computational mechanisms to improve several important algorithm characteristics such as population diversity or searching capacities.

On the other hand, Particle Swarm optimization (PSO) [30] and Artificial Bee Colony (ABC) [31] are the most popular swarm algorithms for solving complex optimization problems. However, they present serious flaws such as premature convergence and difficulty to overcome local minima [32, 33]. Such problems arise from operators that modify individual positions. In such algorithms, the position of each agent in the next iteration is updated yielding an attraction towards the position of the best particle seen so-far (in case of PSO) or any other randomly chosen individual (in case of ABC). Such behaviors produce that the entire population concentrates around the best particle or diverges without control as the algorithm evolves, either favoring the premature convergence or damaging the exploration-exploitation balance [34, 35].

Different to other EA, at SSO each individual is modeled considering the gender. Such fact allows incorporating computational mechanisms to avoid critical flaws such as premature convergence and incorrect exploration-exploitation balance commonly present in both, the PSO and the ABC algorithm. From an optimization point of view, the use of the social-spider behavior as a metaphor introduces interesting concepts in EA: the fact of dividing the entire population into different search-agent categories and the employment of specialized operators that are applied selectively to each of them. By using this framework, it is possible to improve the balance between exploitation and exploration, yet preserving the same population, i.e. individuals who have achieved efficient exploration (female spiders) and individuals that verify extensive exploitation (male spiders). Furthermore, the social-spider behavior mechanism introduces an interesting computational scheme with three important particularities: first, individuals are separately processed according to their characteristics. Second, operators share the same communication mechanism allowing the employment of important information of the evolutionary process to modify the influence of each operator. Third, although operators modify the position of only an individual type, they use global information (positions of all individual types) in order to perform such modification. Figure 4.3 presents a schematic representation of the algorithm-data-flow. According to Fig. 4.3, the female cooperative and male cooperative operators process only female or male individuals, respectively. However, the mating operator modifies both individual types.

4.4 Template Matching Process

Beginning from the problem of locating a given reference image (template) R over a larger intensity image I, the task is to find those positions at image I whose coincident region matches with R or at least is most similar.

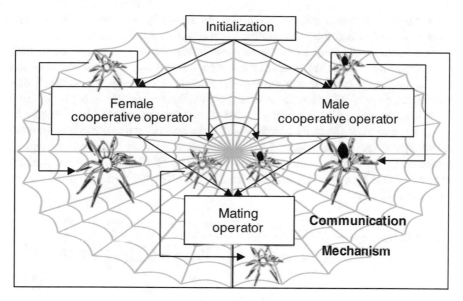

Fig. 4.3 Schematic representation of the SSO algorithm-data-flow

If it is denoted by $R_{u,v}(x, y) = R(x - u, y - v)$, the reference image R is shifted by the distance (u, v) towards the horizontal and vertical directions, respectively. Then the matching problem that is illustrated by Fig. 4.4 can be summarized as follows: considering the source image I and the reference image R, find the offset

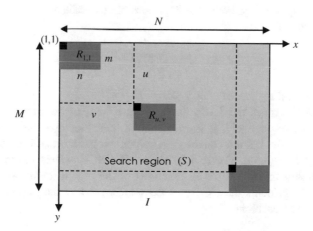

Fig. 4.4 Template matching geometry: the reference image R is shifted across the search image I by an offset (u, v) by using the origins of the two images as reference points. The dimensions of the source image ($M \times N$) and the reference image ($m \times n$) determine the maximal search region (S) for the comparison

(u, v) within the search region S such that the similarity between the shifted reference image $R_{u,v}$ and the corresponding sub-image of I is the maximum.

In order to successfully solve this task, several issues need to be addressed such as determining a minimum similarity value to validate that a match has occurred and developing a good search strategy to find, in a fast way, the optimal displacement. Several TM algorithms that use evolutionary approaches [7–10] have been proposed as a search strategy to reduce the number of search positions. In comparison to other similarity criteria, NCC is the most effective and robust method that supports the measurement of the resemblance between R and its coincident region at image I, at each displacement (u, v). The NCC value between a given image I of size $M \times N$ and a template image R of size $m \times n$, at the displacement (u, v), is given by:

$$NCC(u, v) = \frac{\sum_{i=1}^{m} \sum_{j=1}^{n} [I(u+i, v+j) - \bar{I}(u, v)] \cdot [R(i,j) - \bar{R}]}{\left[\sum_{i=1}^{m} \sum_{j=1}^{n} I(u+i, v+j) - \bar{I}(u, v)\right]^{\frac{1}{2}} \cdot \left[\sum_{i=1}^{m} \sum_{j=1}^{n} R(i,j) - \bar{R}\right]^{\frac{1}{2}}}$$

(4.14)

where $\bar{I}(u, v)$ is the grey-scale average intensity of the source-image for the coincident region of template image R whereas $\bar{R}$ is the grey-scale average intensity of the template image. These values are defined as follows:

$$\bar{I}(u, v) = \frac{1}{m \cdot n} \sum_{i=1}^{m} \sum_{j=1}^{n} I(u+i, v+j) \bar{R} = \frac{1}{m \cdot n} \sum_{i=1}^{m} \sum_{j=1}^{m} R(i,j)$$

(4.15)

The point (u, v) that presents the best possible resemblance between R and I is thus defined as follows:

$$(u, v) = \arg \max_{(\hat{u},\hat{v}) \in S} NCC(\hat{u}, \hat{v})$$

(4.16)

where $S = \{(\hat{u}, \hat{v}) | 1 \leq \hat{u} \leq M - m, 1 \leq \hat{v} \leq N - n\}$.

Figure 4.5 illustrates the TM process. Figure 4.5a, b show the source and the template image respectively. It is important to consider that the template image (Fig. 4.5b) is similar but not equal to the coincident pattern that is contained in the source image (Fig. 4.5a). Figure 4.5c shows the NCC values (color-encoded) in form of contours, calculated for all locations of the search region S. On the other hand, Fig. 4.5d presents the NCC surface which exhibits the highly multi-modal nature of the TM problem.

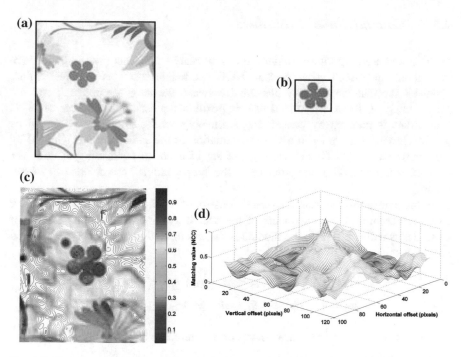

Fig. 4.5 Template matching process. **a** Example source image, **b** template image, **c** color-encoded NCC values and **d** NCC multi-modal surface

4.5 TM Algorithm Based on SSO

4.5.1 Memory Mechanism

The TM problem is defined by using a search space compound by a set of finite solutions. Therefore, since SSO algorithm employs random numbers for the calculation of new individuals, such individuals may encounter the same solutions (repetition) that have been visited by other individuals at previous iterations. Evidently, such fact seriously constraints the SSO performance mainly when the fitness evaluation is computationally expensive to calculate, as it is in the case of the NCC computation.

In order to enhance the performance of the search strategy, the number of NCC evaluations is reduced by considering a memory **FM** which stores the NCC values previously visited in order to avoid re-evaluation of the same particles. The **FM** memory contains a list that includes search position and its corresponding NCC value. Therefore, before evaluating a determined search position, it is analyzed the **FM** memory, if it already contains the search position, then it is not necessary to evaluate it; otherwise the NCC of the search position is computed and stored in the **FM** memory for its later use.

4.5.2   Computational Procedure

In the SSO-based algorithm, spiders represent search positions (u, v) which move throughout the search space S. The NCC coefficient, used as a fitness value, evaluates the matching quality presented between the template image R and the source image I, for a determined search position (spider). The number of NCC evaluations is reduced by considering a memory which stores the NCC values previously visited in order to avoid re-evaluation of the same particles. Guided by the fitness values (NCC coefficients), the set of encoded candidate positions are evolved using the SSO operators until the best possible resemblance has been found.

In the algorithm, the search space S consists of a set of 2-D search positions $\hat{u}$ and $\hat{v}$ representing the components of the search location. Each spider female or male is thus defined as (Fig. 4.6 illustrates this configuration, under the bio-inspired approach):

$$\begin{aligned} \mathbf{f}_i &= \left\{ (\hat{u}_i, \hat{v}_i) \middle| 1 \le \hat{u}_i \le M - m, 1 \le \hat{v}_i \le N - n \right\} \\ \mathbf{m}_j &= \left\{ (\hat{u}_j, \hat{v}_j) \middle| 1 \le \hat{u}_j \le M - m, 1 \le \hat{v}_j \le N - n \right\} \end{aligned} \tag{4.17}$$

The proposed SSO-based algorithm can be summarized in the following steps:

Step 1   Read gray scale image I.
Step 2   Select the template R.
Step 3   Initialize the **FM** memory.
Step 4   Considering N as the total number of n-dimensional colony members, define the number of male N_m and females N_f spiders in the entire population **S**.
Step 5   Initialize randomly the female $(\mathbf{F} = \{\mathbf{f}_1, \mathbf{f}_2, \ldots, \mathbf{f}_{N_f}\})$ and male $(\mathbf{M} = \{\mathbf{m}_1, \mathbf{m}_2, \ldots, \mathbf{m}_{N_m}\})$ members (where $\mathbf{S} = \{\mathbf{s}_1 = \mathbf{f}_1, \mathbf{s}_2 = \mathbf{f}_2, \ldots, \mathbf{s}_{N_f} = \mathbf{f}_{N_f}, \mathbf{s}_{N_f+1} = \mathbf{m}_1, \mathbf{s}_{N_f+2} = \mathbf{m}_2, \ldots, \mathbf{s}_N = \mathbf{m}_{N_m}\}$), calculating also the radius of mating.

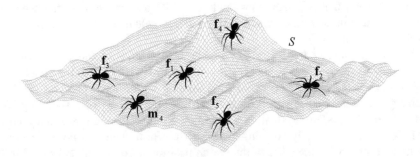

Fig. 4.6 Template matching process under the bio-inspired approach

Step 6   Using the memory mechanism described in Sect. 4.5.1, evaluate the NCC value (Eq. 4.14) for each spider and calculate its respective weight (Sect. 3.1.1).

Step 7   Move female spiders according to the female cooperative operator (Sect. 3.1.4)

Step 8   Move the male spiders according to the male cooperative operator (Sect. 3.1.4).

Step 9   Perform the mating operation (Sect. 3.1.5).

Step 10   Select the best spider which $\mathbf{s}^{Best}$ has the higher NCC value (Eq. 4.14), where $\mathbf{s}^{Best} \in \{\mathbf{S}\}|NCC(\mathbf{s}^{Best}) = \max\{NCC(\mathbf{s}_1), NCC(\mathbf{s}_2), \ldots, NCC(\mathbf{s}_N)\}$.

Step 11   If the number of iterations has been reached, then determine the best individual (matching position) of the final population is $\mathbf{s}^{Best} = \hat{u}_{best}, \hat{v}_{best}$; otherwise go to step 6.

The proposed SSO-TM algorithm considers multiple search locations during the complete optimization process. However, only a few of them are evaluated using the true fitness function whereas all other remaining positions are just taken from the memory **FM**. Figure 4.7 shows a section of the search-pattern that has been generated by the SSO-TM approach considering the problem exposed in Fig. 4.5. Such pattern exhibits the evaluated search-locations in red-cells, whereas the

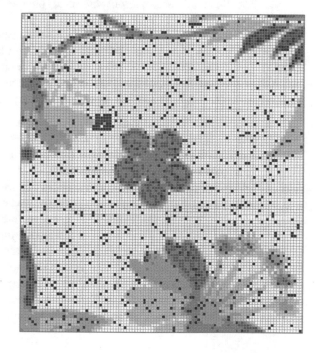

Fig. 4.7 Search-pattern generated by the SSO-TM algorithm. *Green points* represent the evaluated search positions whereas *blue points* indicate the repeated locations. The *red point* exhibits the optimal match detection

maximum location is marked in green. Blue-cells represent those that have been repeated (they were multiply chosen) whereas cells with other gray intensity levels were not visited at all, during the optimization process.

4.6 Experimental Results

In order to verify the feasibility and effectiveness of our proposed algorithm, a series of comparative experiments with other TM algorithms are developed. Simulations have been performed over a set of images that is shown in Fig. 4.8. The proposed approach has been applied to the experimental set and their results have been compared to those produced by the ICA-TM method [14] and the PSO-TM algorithm [12]. Both are considered as state-of-the-art algorithms whose results have been recently published. Notice that the maximum iteration number for the experimental work has been set to 300. Such stop criterion has been selected to maintain compatibility to similar works reported in the literature [11–14].

The parameter set for each algorithm is described as follows:

1. ICA-TM [14]: The parameters are set to: $NumOfCountries = 100$, $NumOfImper = 10$, $NumOf\text{-}Colony = 90$, $T_{max} = 300$, $\xi = 0.1$, $\varepsilon_1 = 0.15$ and $\varepsilon_2 = 0.9$. Such values represent the best parameter set for this algorithm according to [10].
2. PSO-TM [12]: The parameters are set to: particle number = 100, $c_1 = 1.5$ and $c_2 = 1.5$; besides, the particle velocity is initialized between $[-4, 4]$.
3. SSO-TM: After extensive experimentation, the parameter set has been configured as follows: The algorithm has been configured by using: $PF = 0.7$. Once this parameter has been fixed, it is kept for all experiments in this section.

The comparisons are analyzed considering three performance indexes: the average elapsed time (At), the success rate (Sr), the average number of checked locations (AsL) and the average number of function evaluations (AfE). The average elapsed time (At) indicates the time in seconds which is employed during the execution of each single experiment. The success rate (Sr) represents the number of executions in percentage for which the algorithm successfully locates the optimal detection point. The average number of checked locations (AsL) exhibits the number of search locations that have been visited during a single experiment. The average number of function evaluations (AfE) indicates the number of times the NCC coefficient is computed. In order to assure statistical consistency, all these performance indexes are averaged considering a determined number of executions.

Results for 30 runs are reported in Table 4.2. According to this table, SSO-TM delivers better results than ICA and PSO for all images. In particular, the test

Properties	Template	Image

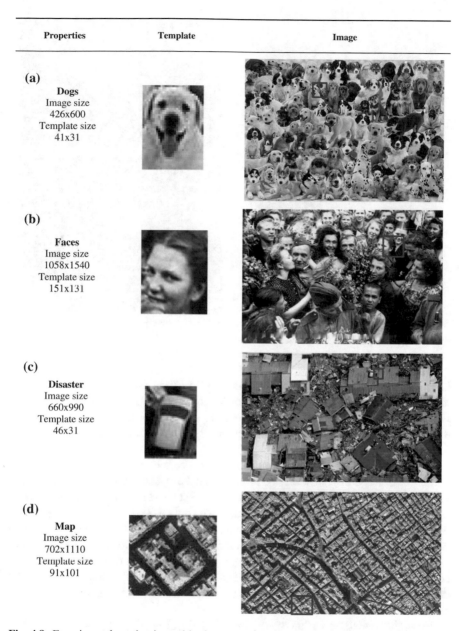

(a)
Dogs
Image size
426x600
Template size
41x31

(b)
Faces
Image size
1058x1540
Template size
151x131

(c)
Disaster
Image size
660x990
Template size
46x31

(d)
Map
Image size
702x1110
Template size
91x101

Fig. 4.8 Experimental set that is used in the comparisons

remarks the largest difference in the success rate (Sr) and the average number of checked locations (AsL). Such facts are directly related to a better trade-off between exploration and exploitation, and the incorporation of the memory mechanism.

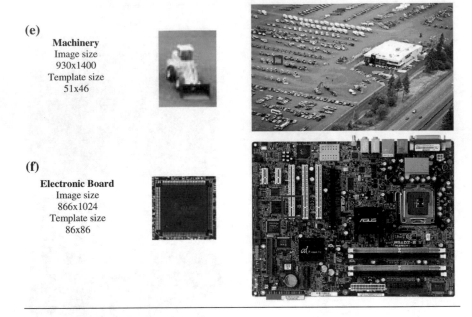

(e)

Machinery
Image size
930x1400
Template size
51x46

(f)

Electronic Board
Image size
866x1024
Template size
86x86

Fig. 4.8 (continued)

In the Table, the exhaustive computation (EC) method (such approach calculates the NCC values for all elements of the source image) is included just as a reference. Finally, Fig. 4.9 presents the matching evolution curve for each image considering the averaged best NCC value seen so-far for all the algorithms that are employed in the comparison.

From Table 4.2, it turns out that the average cost of our algorithm is 5.202 s, while the average cost of the ICA-TM and the PSO-TM algorithms are 5.74 and 7.86 respectively. Such values demonstrate that SSO-TM spends less time on image matching than its counterparts. According to Table 4.2, the SSO-TM presents a better performance than the other two algorithms in terms of effectiveness, since it successfully detects the optimal detection point for all experiments. On the other hand, SSO-TM visits almost the third part of the number of search location in comparison with the ICA-TM and the PSO-TM algorithms. It is important to recall that such evaluation represents the main computational cost which is commonly associated to the TM process.

Recently, statistic tests have been used in several domains [36–41] in order to validate the performance of new approaches over an experimental data set. Therefore, a non-parametric statistical significance proof known as the Wilcoxon's rank sum test for independent samples [42, 43] has been conducted over the average

Table 4.2 Performance comparison of EC, ICA-TM, PSO-TM and the proposed approach for the experimental set shown in Fig. 4.8

Image	Algorithm	Average elapsed time (At)	Success rate (Sr)%	Average number of checked locations (AsL)	Average number of function evaluations (AfE)
a	EC	10.45	100	219,065	219,065
	ICA-TM	1.71	80	30,100	30,100
	PSO-TM	1.55	3.33	29,217	29,217
	SSO-TM	1.35	100	7856	16,521
b	EC	280.89	100	1,277,963	1,277,963
	ICA-TM	7.10	100	30,100	30,100
	PSO-TM	11.02	26.66	29,259	29,259
	SSO-TM	4.19	100	2654	4842
c	EC	28.67	100	588,826	588,826
	ICA-TM	7.01	50	30,100	30,100
	PSO-TM	10.27	20	29,342	29,342
	SSO-TM	6.84	99.66	7896	18,677
d	NCC	68.31	100	616,499	616,499
	ICA-TM	7.17	36.66	30,100	30,100
	PSO-TM	9.82	36.66	29,369	29,369
	SSO-TM	7.14	96.66	10,561	18,933
e	EC	70.20	100	1,190,166	1,190,166
	ICA-TM	4.02	87.66	30,100	30,100
	PSO-TM	4.00	3.33	29,295	29,295
	SSO-TM	4.35	99.66	11,398	19,796
f	EC	74.98	100	731,640	731,640
	ICA-TM	7.45	56.66	30,100	30,100
	PSO-TM	10.53	0	29,116	29,116
	SSO-TM	7.36	100	9296	17,633

number of function evaluations (AfE) data of Table 4.2, with an 5% significance level. Table 4.3 reports the p-values produced by Wilcoxon's test for the pair-wise comparison of the average number of function evaluations (AfE) of four groups. Such groups are constituted by SSO-TM versus ICA-TM and SSO versus PSO-TM. As a null hypothesis, it is assumed that there is no significant difference between mean values of the two algorithms. The alternative hypothesis considers a significant difference between the AfE values of both approaches. All p-values reported in Table 4.3 are less than 0.05 (5% significance level) which is a strong evidence against the null hypothesis. Therefore, such evidence indicates that SSO-TM results are statistically significant and that it has not occurred by coincidence (i.e. due to common noise contained in the process).

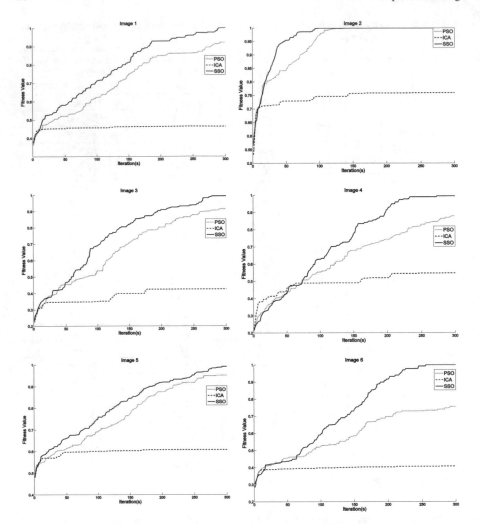

Fig. 4.9 Evolution curves for ICA-TM, PSO-TM and the proposed SSO-TM considering the average best NCC value seen so-far, each curve corresponds to the image of the experimental set

Table 4.3 *p*-values produced by Wilcoxon's test comparing SSO-TM versus ICA-TM and SSO versus PSO-TM over the average number of function evaluations (AfE) values from Table 4.2

Image	SSO-TM versus ICA-TM	SSO-TM versus PSO-TM
(a)	1.12E−10	1.18E−10
(b)	2.34E−12	4.64E−12
(c)	0.98E−12	2.95E−12
(d)	1.38E−12	3.35E−12
(e)	2.21E−12	5.10E−12
(f)	1.09E−12	3.37E−12

4.7 Conclusions

In this chapter, a novel bio-inspired algorithm called the Social Spider Optimization (SSO) is proposed to reduce the number of search locations in the TM process. The SSO algorithm is based on the simulation of cooperative behavior of social-spiders. The algorithm considers two different search individuals (spiders): males and females. Depending on gender, each individual is conducted by a set of different evolutionary operators which mimic different cooperative behaviors that are typically found in the colony.

In the proposed approach, spiders represent search locations which move throughout the positions of the source image. The NCC coefficient, used as a fitness value, evaluates the matching quality presented between the template image and the coincident region of the source image, for a determined search position (spider). The number of NCC evaluations is reduced by considering a memory which stores the NCC values previously visited in order to avoid the re-evaluation of the same search locations. Guided by the fitness values (NCC coefficients), the set of encoded candidate positions are evolved using the SSO operators until the best possible resemblance has been found.

The performance of the proposed approach has been compared to other existing TM algorithms by considering different images that show a great variety of formats and complexities. Experimental results demonstrate a higher performance of the proposed method in terms of the elapsed time and the number of NCC evaluations.

References

1. Brabazon, A., O'Neill, M.: Biologically Inspired Algorithms for financial Modelling. Srpinger, Berlin (2006).
2. Gordon, D., The Organization of Work in Social Insect Colonies, Complexity, 8(1), (2003), 43–46.
3. Lubin, T. B. The Evolution of Sociality in Spiders. In H. J. Brockmann, Advances in the study of behavior, 37, (2007), 83–145.
4. Uetz, G. W. Colonial web-building spiders: Balancing the costs and. In E. J. Choe and B. Crespi, The Evolution of Social Behavior in Insects and Arachnids (pp. 458–475). Cambridge, England.: Cambridge University Press.
5. Aviles, L. Sex-Ratio Bias and Possible Group Selection in the Social Spider Anelosimus eximius. The American Naturalist, 128(1), (1986), 1–12.
6. Burgess, J. W. Social spacing strategies in spiders. In P. N. Rovner, Spider Communication: Mechanisms and Ecological Significance (pp. 317–351). Princeton, New Jersey.: Princeton University Press, (1982).
7. Maxence, S. Social organization of the colonial spider Leucauge sp. in the Neotropics: vertical stratification within colonies. The Journal of Arachnology 38, (2010), 446–451.
8. Eric C. Yip, K. S. Cooperative capture of large prey solves scaling challenge faced by spider societies. Proceedings of the National Academy of Sciences of the United States of America, 105(33), (2008), 11818–11822.

9. Hadi, G., Mojtaba, L., Hadi, S.Y., 2009. An improved pattern matching technique for lossy/lossless compression of binary printed Farsi and Arabic textual images. Int. J. Intell. Comput. Cybernet. 2 (1), 120–147.

10. Krattenthaler, W., Mayer, K.J., Zeiler, M., 1994. Point correlation: A reduced-cost template matching technique. In: Proceedings of the First IEEE International Conference on Image Processing, pp. 208–212.

11. Na Dong, Chun-Ho Wu, Wai-Hung Ip, Zeng-Qiang Chen, Ching-Yuen Chan, Kai-Leung Yung. An improved species based genetic algorithm and its application in multiple template matching for embroidered pattern inspection, Expert Systems with Applications, 38, (2011), 15172–15182.

12. Fang Liu, Haibin Duana, Yimin Deng. A chaotic quantum-behaved particle swarm optimization based on lateral inhibition for image matching, Optik, 123, (2012), 1955–1960.

13. Chun-Ho Wu, Da-Zhi Wang, Andrew Ip, Ding-Wei Wang, Ching-Yuen Chan, Hong-Feng Wang. A particle swarm optimization approach for components placement inspection on printed circuit boards, J Intell Manuf 20, (2009), 535–549.

14. Haibin Duan, Chunfang Xu, Senqi Liu, Shan Shao. Template matching using chaotic imperialist competitive algorithm, Pattern Recognition Letters, 31, (2010), 1868–1875.

15. Chen, G., Low, C.P., Yang, Z. Preserving and exploiting genetic diversity in evolutionary programming algorithms. IEEE Transactions on Evolutionary Computation 13(3), (2009), 661–673.

16. Adra, S.F., Fleming, P.J. Diversity management in evolutionary many-objective optimization. IEEE Transactions on Evolutionary Computation 15(2), (2011), 183–195.

17. Oster, G., Wilson, E. Caste and ecology in the social insects. Princeton, N.J. Princeton University press, 1978.

18. Bert Hölldobler, E.O. Wilson. Journey to the Ants: A Story of Scientific Exploration, 1994, ISBN 0-674-48525-4.

19. Bert Hölldobler, E.O. Wilson: "The Ants, Harvard University Press, 1990, ISBN 0-674-04075-9.

20. Avilés, L. Causes and consequences of cooperation and permanent-sociality in spiders. In B. C. Choe., The Evolution of Social Behavior in Insects and Arachnids (pp. 476–498). Cambridge, Massachusetts.: Cambridge University Press, 1997.

21. Rayor, E. C. Do social spiders cooperate in predator defense and foraging without a web? Behavioral Ecology & Sociobiology, 65(10), 2011, 1935–1945.

22. Gove, R., Hayworth, M., Chhetri, M., Rueppell, O. Division of labour and social insect colony performance in relation to task and mating number under two alternative response threshold models, Insect. Soc. 56(3), (2009), 19–331.

23. Ann L. Rypstra, R. S. Prey Size, Prey Perishability and Group Foraging in a Social Spider. Oecologia 86, (1), (1991), 25–30.

24. Pasquet, A. Cooperation and prey capture efficiency in a social spider, Anelosimus eximius (Araneae, Theridiidae). Ethology 90, (1991), 121–133.

25. Ulbrich, K., Henschel, J. Intraspecific competition in a social spider, Ecological Modelling, 115(2–3), (1999), 243–251.

26. Jones, T., Riechert, S. Patterns of reproductive success associated with social structure and microclimate in a spider system, Animal Behaviour, 76(6), (2008), 2011–2019.

27. Damian O., Andrade, M., Kasumovic, M. Dynamic Population Structure and the Evolution of Spider Mating Systems, Advances in Insect Physiology, 41, (2011), 65–114.

28. Yang X-S (2008) Nature-inspired metaheuristic algorithms. Luniver Press, Beckington.

29. Chen DB, Zhao CX (2009) Particle swarm optimization with adaptive population size and its application. Appl Soft Comput 9(1):39–48.

30. J. Kennedy and R. Eberhart, Particle swarm optimization, in Proceedings of the 1995 IEEE International Conference on Neural Networks, vol. 4, pp. 1942–1948, December 1995.

31. Karaboga, D. An Idea Based on Honey Bee Swarm for Numerical Optimization. Technical Report-TR06. Engineering Faculty, Computer Engineering Department, Erciyes University, 2005.

32. Wang, Y., Li, B., Weise, T., Wang, J., Yuan, B., Tian, Q. Self-adaptive learning based particle swarm optimization, Information Sciences, 181(20), (2011), 4515–4538.
33. Wan-li, X., Mei-qing A. An efficient and robust artificial bee colony algorithm for numerical optimization, Computers & Operations Research, 40, (2013), 1256–1265.
34. Wang, H., Sun, H., Li, C., Rahnamayan, S., Jeng-shyang P. Diversity enhanced particle swarm optimization with neighborhood, Information Sciences, 223, (2013), 119–135.
35. Banharnsakun, A., Achalakul, T., Sirinaovakul, B. The best-so-far selection in Artificial Bee Colony algorithm, Applied Soft Computing 11, (2011), 2888–2901.
36. Oliva, D., Cuevas, E., Pajares, G., Zaldivar, D., Osuna, V., A Multilevel thresholding algorithm using electromagnetism optimization, Neurocomputing, (2014), 357–381.
37. Oliva, D., Cuevas, E., Pajares, G., Zaldivar, D., Perez-Cisneros, M., Multilevel thresholding segmentation based on harmony search optimization, Journal of Applied Mathematics, 2013, 575414.
38. Cuevas, E., Zaldivar, D., Pérez-Cisneros, M., Seeking multi-thresholds for image segmentation with Learning Automata, Machine Vision and Applications, 22 (5), (2011), 805–818.
39. Cuevas, E., Ortega-Sánchez, N., Zaldivar, D., Pérez-Cisneros, M., Circle detection by Harmony Search Optimization, Journal of Intelligent and Robotic Systems: Theory and Applications, 66 (3), (2012), 359–376.
40. Cuevas, E., Zaldivar, D., Pérez-Cisneros, M., Ramírez-Ortegón, M., Circle detection using discrete differential evolution Optimization, Pattern Analysis and Applications, 14 (1), (2011), 93–107.
41. Cuevas, E., Echavarría, A., Zaldívar, D., Pérez-Cisneros, M., A novel evolutionary algorithm inspired by the states of matter for template matching, Expert Systems with Applications, 40 (16), (2013), 6359–6373
42. Wilcoxon F (1945) Individual comparisons by ranking methods. Biometrics 1:80–83.
43. Garcia S, Molina D, Lozano M, Herrera F (2008) A study on the use of non-parametric tests for analyzing the evolutionary algorithms' behaviour: a case study on the CEC'2005 Special session on real parameter optimization. J Heurist. doi:10.1007/s10732-008-9080-4.

Chapter 5
Motion Estimation

Abstract Motion estimation is a major problem for video-coding applications. Among several other motion estimation approaches, block matching (BM) algorithms are the most popular methods due to their effectiveness and simplicity at their software and hardware implementation. The BM approach assumes that the pixel movement inside a given region of the current frame (Macro-Block, MB) can be modeled as a pixel translation from its corresponding region in the previous frame. In this procedure, the motion vector is obtained by minimizing the sum of absolute differences (SAD) from the current frame's MB over a determined search window from the previous frame. Unfortunately, the SAD evaluation is computationally expensive and represents the most consuming operation in the BM process. The simplest available BM method is the full search algorithm (FSA) which finds the most accurate motion vector through an exhaustive computation of SAD values for all elements of the search window. However, several fast BM algorithms have been lately presented to reduce the number of SAD operations by calculating only a fixed subset of search locations at the price of poor accuracy. In this chapter, a new algorithm based on Differential Evolution (DE) is presented to reduce the number of search locations in the BM process. In order to avoid the computing of several search locations, the algorithm estimates the SAD (fitness) values for some locations by considering SAD values from previously calculated neighboring positions. Since the presented algorithm does not consider any fixed search pattern or other different assumption, a high probability for finding the true minimum (accurate motion vector) is expected. In comparison to other fast BM algorithms, the presented method deploys more accurate motion vectors yet delivering competitive time rates.

5.1 Introduction

Virtually all video applications and visual communications deal with an enormous amount of data. The limited storage capacity and the available transmission bandwidth both have made digital video coding an important technology. Its high correlation between successive frames can be exploited to improve coding efficiency

© Springer International Publishing AG 2017   95
E. Cuevas et al., *Evolutionary Computation Techniques:*
A Comparative Perspective, Studies in Computational Intelligence 686,
DOI 10.1007/978-3-319-51109-2_5

which is usually achieved by using motion estimation (ME). Several ME methods have been studied aiming for a complexity reduction at video coding, such as block matching (BM) algorithms, parametric-based models [1], optical flow [2] and pel-recursive techniques [3]. Among such methods, BM seems to be the most popular technique due to its effectiveness and simplicity for both software and hardware implementations. BM is also widely adopted by various video coding standards such as MPEG-1 [4], MPEG-2 [5], MPEG-4 [6], H.261 [7] and H.263 [8].

For BM algorithms, the current frame is divided into non-overlapping macro blocks of $N \times N$ pixel dimension. For each block in the current frame, the best matched block within a search window of size $(2W + 1) \times (2W + 1)$ is determined at the previous frame, where W is the maximum allowed displacement. The position difference between a template block in the current frame and the best matched block in the previous frame is called the motion vector (MV). A commonly used matching measure is the sum of absolute differences (SAD) which is computationally expensive and represents the most consuming operation in the BM process.

The full search algorithm (FSA) [9] is the simplest block-matching algorithm that can deliver the optimal estimation solution regarding a minimal matching error as it checks all candidates one at a time. However, such exhaustive search and full-matching error calculation at each checking point yields an extremely computational expensive BM method that seriously constraints real-time video applications.

In order to decrease the computational complexity of the BM process, several BM algorithms have been presented considering the following three techniques: (1) using a fixed pattern: which means that the search operation is conducted over a fixed subset of the total search window. The Three Step Search (TSS) [10], the New Three Step Search (NTSS) [11], the Simple and Efficient TSS (SES) [12], the Four Step Search (4SS) [13] and the Diamond Search (DS) [14] are some of its well-known examples. Although such approaches have been algorithmically considered as the fastest, they are not able eventually to match the dynamic motion-content, delivering false motion vectors (image distortions). (2) Reducing the search points: in this method, the algorithm chooses as search points exclusively those locations which iteratively minimize the error-function (SAD values). This category includes: the Adaptive Rood Pattern Search (ARPS) [15], the Fast Block Matching Using Prediction (FBMAUPR) [16], the Block-based Gradient Descent Search (BBGD) [17] and the Neighbourhood Elimination algorithm (NE) [18]. Such approaches assume that the error-function behaves monotonically, holding well for slow-moving sequences; however, such properties do not hold true for other kind of movements in video sequences [19], which risks on algorithms getting trapped into local minima. (3) Decreasing the computational overhead for every search point, which means the matching cost (SAD operation) is replaced by a partial or a simplify version that features less complexity. The New pixel-Decimation (ND) [20], the Efficient Block Matching Using Multilevel Intra and Inter-Sub-blocks [11] and the Successive Elimination Algorithm [21], all assume that all pixels within each block move by the same amount and a good estimate of the motion could be obtained through only a fraction of the pixel pool. However, since only a fraction of pixels enters into the matching computation, the use of these regular sub-sampling techniques can seriously affect the accuracy of the detection of motion vectors due to noise or illumination changes.

Alternatively, evolutionary approaches such as genetic algorithms (GA) [22] and particle swarm optimization (PSO) [23] are well known for locating potential global optimum within an arbitrary search space. Despite of such fact, only few evolutionary approaches have specifically addressed the problem of BM, such as the light-weight genetic block matching (LWG) [24], the genetic four-step search (GFSS) [25] and the PSO-BM [26]. Although these methods support an accurate identification of the motion vector, their spending times are very long in comparison to other BM techniques.

Differential Evolution (DE), introduced by Storn and Price in 1995 [27], is a novel evolutionary algorithm which is used to optimize complex continuous non-linear functions. As a population-based algorithm, DE uses simple mutation and crossover operators to generate new candidate solutions. It also applies one-to-one competition schemes to greedily decide whether the new candidate or its parent will survive in the next generation. Due to its simplicity, ease of implementation, fast convergence and robustness, the DE algorithm has gained much attention that has yielded a wide range of successful applications in the literature [28–36].

For many real-world applications, the number of calls to the objective function needs to be limited because an evaluation is either time consuming, expensive or requires user interaction. DE does not seem to be suited to face such problems as it usually requires many evaluations before delivering an acceptable result.

The problem of excessively long fitness function calculations has already been faced in the field of evolutionary algorithms (EA) and is better known as evolution control [37]. For such approach, the idea is to replace the costly objective function evaluation for some individuals by fitness estimates which are based on an approximate model of the fitness landscape. The individuals to be evaluated and those to be estimated are determined following some fixed criteria which depend on the specific properties of the approximate model [38]. The models involved at the estimation can be built during the actual EA run, since EA repeatedly sample the search space at different points [39]. There are many possible approximation models and several have already been used in combination with EA (e.g. polynomials [40], the kriging model [41], the feed-forward neural networks that includes multi-layer Perceptrons [42] and radial basis-function networks [43]). These models can be either global, which make use of all available data or local which make use of only a small set of data around the point where the function is to be approximated. Local models, however, have a number of advantages [39]: they are well-known and suitably established techniques with relatively fast speeds. Moreover, they consider the intuitively most important information: the closest neighbors.

In this chapter, a new algorithm based on Differential Evolution (DE) is presented to reduce the number of search locations in the BM process. The algorithm uses a simple fitness calculation approach which is based on the Nearest Neighbor

Interpolation (NNI) algorithm in order to estimate the fitness value (SAD operation) for several candidate solutions (search locations). As a result, the approach can substantially reduce the number of function evaluations preserving the good search capabilities of DE. In comparison to other fast BM algorithms, the presented method deploys more accurate motion vectors yet delivering competitive time rates.

The overall chapter is organized as follows: Sect. 5.2 holds a brief description about the differential evolution algorithm. In Sect. 5.3, the fitness calculation strategy for solving the expensive optimization problem is presented. Section 5.4 provides background about the BM motion estimation issue while Sect. 5.5 exposes the final BM algorithm as a combination of DE and the NNI estimator. Section 5.6 demonstrates experimental results for the presented approach over standard test sequences and some conclusions are drawn in Sect. 5.7.

5.2 Differential Evolution Algorithm

The DE algorithm is a simple and direct search algorithm which is based on population and aims for optimizing global multi-modal functions. DE employs the mutation operator as to provide the exchange of information among several solutions.

There are various mutation base generators to define the algorithm type. The version of DE algorithm used in this work is known as DE/best/l/exp or "DE1" [27]. DE algorithms begin by initializing a population of N_p and D-dimensional vectors considering parameter values that are randomly distributed between the pre-specified lower initial parameter bound $x_{j,\mathrm{low}}$ and the upper initial parameter bound $x_{j,\mathrm{high}}$ as follows:

$$x_{j,i,t} = x_{j,\mathrm{low}} + \mathrm{rand}(0,1) \cdot (x_{j,\mathrm{high}} - x_{j,\mathrm{low}}); \tag{5.1}$$
$$j = 1,2,\ldots,D; \quad i = 1,2,\ldots,N_p; \quad t = 0.$$

The subscript t is the generation index, while j and i are the parameter and particle indexes respectively. Hence, $x_{j,i,t}$ is the j-th parameter of the i-th particle in generation t. In order to generate a trial solution, DE algorithm first mutates the best solution vector $\mathbf{x}_{best,t}$ from the current population by adding the scaled difference of two vectors from the current population.

$$\mathbf{v}_{i,t} = \mathbf{x}_{best,t} + F \cdot (\mathbf{x}_{r_1,t} - \mathbf{x}_{r_2,t}); \tag{5.2}$$
$$r_1, r_2 \in \{1,2,\ldots,N_p\}$$

with $\mathbf{v}_{i,t}$ being the mutant vector. Indexes r_1 and r_2 are randomly selected with the condition that they are different and have no relation to the particle index i whatsoever (i.e., $r_1 \neq r_2 \neq i$). The mutation scale factor F is a positive real number, typically less than one. Figure 5.1 illustrates the vector-generation process defined by Eq. (5.2).

Fig. 5.1 Two-dimensional example of an objective function showing its contour lines and the process for generating v in scheme DE/best/l/exp from current generation's vectors

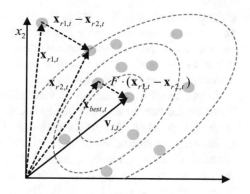

In order to increase the diversity of the parameter vector, the crossover operation is applied between the mutant vector and the original individuals. The result is the trial vector is computed by considering element to element as follows:

$$u_{j,i,t} = \begin{cases} v_{j,i,t}, & \text{if } \text{rand}(0,1) \leq CR \quad \text{or} \quad j = j_{\text{rand}}, \\ x_{j,i,t}, & \text{otherwise.} \end{cases} \qquad (5.3)$$

with $j_{\text{rand}} \in \{1, 2, \ldots, D\}$. The crossover parameter $(0.0 \leq CR \leq 1.0)$ controls the fraction of parameters that the mutant vector is contributing to the final trial vector. In addition, the trial vector always inherits the mutant vector parameter according to the randomly chosen index j_{rand}, assuring that the trial vector differs by at least one parameter from the vector to which it is compared $(x_{i,t})$.

Finally, a greedy selection is used to find better solutions. Thus, if the computed cost function value of the trial vector $\mathbf{u}_{i,t}$ is less or equal than the cost of the vector $\mathbf{x}_{i,t}$, then such trial vector replaces $\mathbf{x}_{i,t}$ in the next generation. Otherwise, $\mathbf{x}_{i,t}$ remains in the population for at least one more generation:

$$\mathbf{x}_{i,t+1} = \begin{cases} \mathbf{u}_{i,t}, & \text{if } f(\mathbf{u}_{i,t}) \leq f(\mathbf{x}_{i,t}), \\ \mathbf{x}_{i,t}, & \text{otherwise.} \end{cases} \qquad (5.4)$$

Here, $f(\ )$ represents the cost function. These processes are repeated until a termination criterion is attained or a predetermined generation number is reached.

5.3 Fitness Approximation Method

Evolutionary algorithms based on fitness approximation aim to find the global minimum of a given function considering only a very few number of function evaluations. In order to apply such approach, it is necessary that the objective function holds the following conditions [44]: (1) its evaluation must be very costly and (2) it must have few dimensions (up to five). Recently, several fitness estimators have been reported in the literature [40–43], with a function evaluation number which is considerably reduced to hundreds, dozens, or even less. However, most of these methods produce complex algorithms whose performance is conditioned to the quality of the training phase and the learning algorithm in the construction of the approximation model.

In this chapter, we explore the use of a local approximation scheme which is based on the nearest-neighbor-interpolation (NNI) and aims to reduce the function evaluation number. The model estimates the fitness values based on previously evaluated neighboring individuals which have been stored during the evolution process. At each generation, some individuals of the population are evaluated through the accurate (real) objective function while other remaining individual fitness values are only estimated. The positions to be accurately evaluated are determined based on their proximity to the best individual or regarding their uncertain fitness value.

5.3.1   Updating the Individual Database

In our fitness calculation approach, every evaluation or estimation of an individual produces a data point (individual position and fitness value) that is potentially taken into account for building the approximation model during the evolution process. Therefore, we keep all seen-so-far evaluations in a history array **T** which is employed to select the closest neighbor and to estimate the fitness value of a new individual. Since all data are preserved and potentially available for use, the model construction is fast because only most relevant data points are actually used by the approach.

5.3.2   Fitness Calculation Strategy

The presented fitness calculation scheme estimates most of fitness values to reduce the calculation time at each generation. In the model, those individuals near the individual with the best fitness value stored at array **T** (rule 1), are evaluated by using the real fitness function. Such individuals are important as they possess a stronger influence on the evolution process than the others. Moreover, it is also evaluated those individuals lying in regions of the search space with few previous evaluations (rule 2). The fitness values of these individuals are uncertain since there is no close reference (close points contained in **T**) in order to calculate their estimates.

The rest of the individuals are estimated using NNI (rule 3). Thus, the fitness value of an individual is estimated assigning it the same fitness value that the nearest individual stored in **T**.

Therefore, the estimation model follows three important rules to evaluate or estimate fitness values:

1. **Exploitation rule (evaluation).** If the new individual (search position) P is located closer than a distance d with respect to the nearest individual location L_q whose fitness value F_{L_q} corresponds to the best fitness value stored in **T**, then the fitness value of P is evaluated using the real fitness function. Figure 5.2a draws the rule procedure.
2. **Exploration rule (evaluation).** If the new individual P is located longer than a distance d with respect to the nearest individual location L_q whose fitness value F_{L_q} has been already stored in **T**, then its fitness value is evaluated using the real fitness function. Figure 5.2b outlines the rule procedure.
3. **NNI rule (estimation).** If the new individual P is located closer than a distance d with respect to the nearest individual location L_q whose fitness value F_{L_q} has been already stored in **T**, then its fitness value is estimated (using the NNI approach) assigning it the same fitness that $L_q(F_P = F_{L_q})$. Figure 5.2c sketches the rule procedure.

The d value controls the trade-off between the evaluation and the estimation of search locations. Typical values of d range from 1 to 4; however, in this chapter, the value of 2.5 has been selected. The presented method, from an optimization perspective, favors the exploitation and exploration in the search process. For the exploration, the method evaluates the fitness function of new search locations which have been located far from previously calculated positions. Additionally, it also estimates those that are closer. For the exploitation, the presented method evaluates the effective fitness function of those new searching locations which are located nearby the position that holds the minimum fitness value seen so far, aiming to improve its minimum.

The three rules show that the fitness calculation strategy is simple and straightforward. Figure 5.2 illustrates the procedure of fitness computation for a new solution (point P). In the problem, the objective function f is minimized with respect to two parameters (x_1, x_2). In all figures, the individual database array T contains five different elements $(L_1, L_2, L_3, L_4, L_5)$ with their corresponding fitness values $(F_{L_1}, F_{L_2}, F_{L_3}, F_{L_4}, F_{L_5})$. Figures 5.2a, b show the fitness evaluation $(f(x_1, x_2))$ of the new solution P, following the rule 1 and 2 respectively, whereas Fig. 5.2c present the fitness estimation of P using the NNI approach which is laid by rule 3.

5.3.3 A New Optimization Method DE

In this section, a fitness calculation approach is presented to accelerate the DE algorithm. Only the fitness calculation scheme shows the difference between the conventional DE and the enhanced approach. In the modified DE, only some

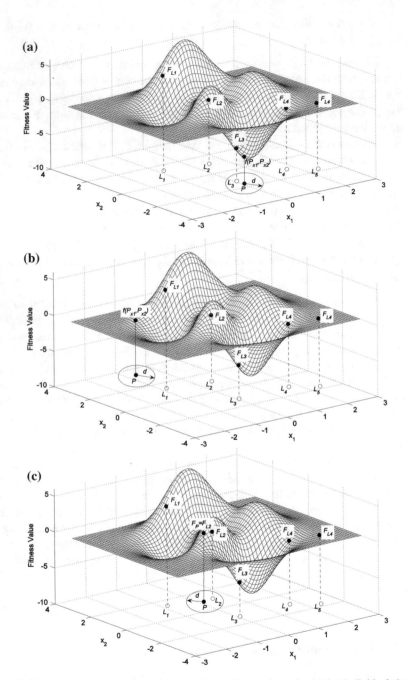

Fig. 5.2 The fitness calculation strategy. **a** According to the rule 1, the individual (search position) P is evaluated since it is located closer than a distance d with respect to the nearest individual location. Therefore, the fitness value corresponds to the best fitness value (minimum). **b** According to the rule 2, the search point P is evaluated and there is no close reference within its neighborhood. **c** According to rule 3, the fitness value of P is estimated by means of the NNI-estimator, assigning

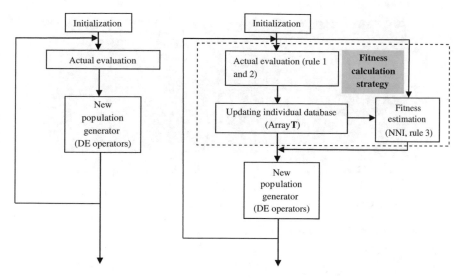

Fig. 5.3 Differences between the conventional DE and the modified DE. **a** Conventional DE and **b** the DE algorithm including the fitness calculation strategy

individuals are actually evaluated (rules 1 and 2) at each generation. The fitness values for the rest are estimated using the NNI-approach (rule 3). The estimation is executed using an individual database (array **T**).

Figure 5.3 shows the difference between the conventional DE and its modified version. It is clear that two new blocks have been added, the fitness estimation and the updating individual database. Both elements and the actual evolution block represent the fitness calculation strategy that is explained at this section. As a result, the DE approach can substantially reduce the number of function evaluations preserving the good search capabilities of DE.

5.4 Motion Estimation and Block Matching

For motion estimation, in a BM algorithm, the current frame of an image sequence I_t is divided into non-overlapping blocks of $N \times N$ pixels. For each template block in the current frame, the best matched block within a search window (S) of size $(2W + 1) \times (2W + 1)$ in the previous frame I_{t-1} is determined, where W is the maximum allowed displacement. The position difference between a template block in the current frame and the best matched block in the previous frame is called the Motion Vector (MV) (see Fig. 5.4).

The most well-known criterion for BM algorithms is the sum of absolute difference (SAD). It is defined in Eq. (5.5) considering a template block at position (x, y) in the current frame and the candidate block at position $(x + \hat{u}, y + \hat{v})$ in the previous frame I_{t-1}.

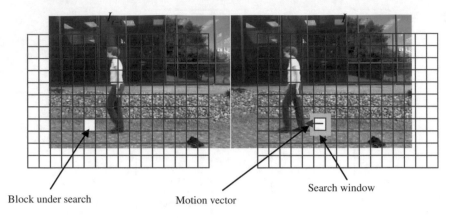

Block under search   Motion vector

Search window

Fig. 5.4 Block matching procedure

$$\text{SAD}(\hat{u}, \hat{v}) = \sum_{j=0}^{N-1} \sum_{i=0}^{N-1} |g_t(x+i, y+j) - g_{t-1}(x+\hat{u}+i, y+\hat{v}+j)| \qquad (5.5)$$

where $g_t(\cdot)$ is the gray value of a pixel in the current frame I_t and $g_{t-1}(\cdot)$ is the gray level of a pixel in the previous frame I_{t-1}. Therefore, the MV in (u, v) is defined as follows:

$$(u, v) = \arg \min_{(u,v) \in S} \text{SAD}(\hat{u}, \hat{v}) \qquad (5.6)$$

where $S = \{(\hat{u}, \hat{v}) | -W \le \hat{u}, \hat{v} \le W \text{ and } (x+\hat{u}, y+\hat{v}) \text{ is a valid pixel position } I_{t-1}\}$.

In the context of BM algorithms, the FSA is the most robust and accurate method to find the MV. It tests all possible candidate blocks from I_{t-1} within the search area to find the block with minimum SAD. For the maximum displacement of W, the FSA requires $(2W+1)^2$ search points. For instance, if the maximum displacement W is 7, the total search points are 225. Each SAD calculation requires $2N^2$ additions and the total number of additions for the FSA to match a 16×16 block is 130,560. Such computational requirement makes the application of FSA difficult for real time applications.

5.5 BM Algorithm Based on DE with the Estimation Strategy

FSA finds the global minimum (the accurate MV), considering all locations within the search space S. Nevertheless, the approach has a high computational cost for practical use. In order to overcome such a problem, many fast algorithms have been

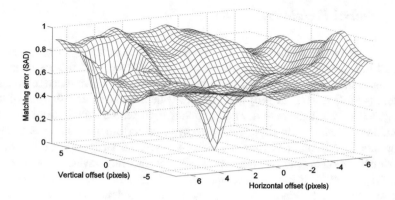

Fig. 5.5 Common non-uni-modal error surface with multiple local minimum error points

developed yielding only a poorer precision than the FSA. A better BM algorithm should spend less computational time on searching and obtaining accurate motion vectors (MVs).

The BM algorithm presented at this chapter is comparable to the fastest algorithms and delivers a similar precision to the FSA approach. Since most of fast algorithms use a regular search pattern or assume a characteristic error function (uni-modal) for searching the motion vector, they may get trapped into local minima considering that for many cases (i.e. complex motion sequences) an uni-modal error is no longer valid. Figure 5.5 shows a typical error surface (SAD values) which has been computed around the search window for a fast-moving sequence. On the other hand, the presented BM algorithm uses a non-uniform search pattern for locating global minimum distortion. Under the effect of the DE operators, the search locations vary from generation to generation, avoiding to get trapped into a local minimum. Besides, since the presented algorithm uses a fitness calculation strategy for reducing the evaluation of the SAD values, it requires fewer search positions.

In the algorithm, the search space S consists of a set of 2-D motion vectors $\hat{u}$ and $\hat{v}$ representing the x and y components of the motion vector, respectively. The particle is defined as:

$$P_i = \{\hat{u}_i, \hat{v}_i | - W \leq \hat{u}_i, \hat{v}_i \leq W\} \tag{5.7}$$

where each particle i represents a possible motion vector. In this chapter, the maximum offset is $W = 7$ pixels.

5.5.1 Initial Population

The first step in DE optimization is to generate an initial group of particles. The standard literature of evolutionary algorithms generally suggests the use of random solutions as the initial population, considering the absence of knowledge about the problem [45]. However, Li [46] and Xiao [47] demonstrated that the use of solutions generated through some domain knowledge to set the initial population (i.e., non-random solutions) can significantly improve its performance. In order to obtain appropriate initial solutions (based on knowledge), an analysis over the motion vector distribution should be conducted. After considering several sequences (see Table 5.1; Fig. 5.9), it can be seen that 98% of the MVs are found to lie at the origin of the search window for a slow-moving sequence such as the one at *Container*, whereas complex motion sequences, such as the *Carphone* and the *Foreman* examples, have only 53.5 and 46.7% of their MVs in the central search region. The *Stefan* sequence, showing the most complex motion content, has only 36.9%. Figure 5.6 shows the surface of the MV distribution for the *Foreman* and the *Stefan*. On the other hand, although it is less evident, the MV distribution of several sequences shows small peaks at some locations lying away from the center as they are contained inside a rectangle that is shown in Fig. 5.6b, d by a white overlay. Real-world moving sequences concentrate most of the MVs under a limit due to the motion continuity principle [41]. Therefore, in this chapter, initial solutions are selected from five fixed locations which represent points showing the higher concentration in the MV distribution, just as it is shown by Fig. 5.7.

5.5.2 The DE-BM Algorithm

The goal of our BM-approach is to reduce the number of evaluations of the SAD values (real fitness function) avoiding any performance loss and achieving an acceptable solution. The DE-BM method is listed below:

Table 5.1 Test sequences used in the comparison test

Sequence	Format	Total frames	Motion type
Container	QCIF(176 × 14)	299	Low
Carphone	QCIF(176 × 14)	381	Medium
Foreman	QCIF(352 × 28)	398	Medium
Akiyo	QCIF(352 × 28)	211	Medium
Stefan	CIF(352 × 288)	89	High
Tennis	SIF(352 × 240)	150	High

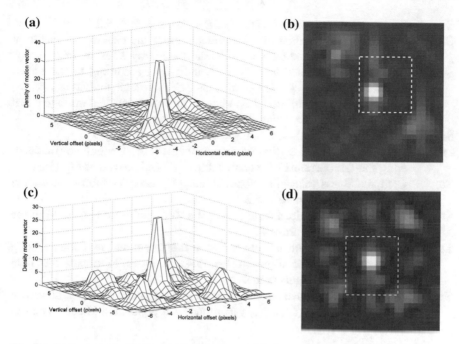

Fig. 5.6 Motion vector distribution for *Foreman* and Stefan sequences. **a–b** MV distribution for the *Foreman* sequence. **c–d** MV distribution for the *Stefan* sequence

Fig. 5.7 Fixed pattern of five elements used as initial solutions

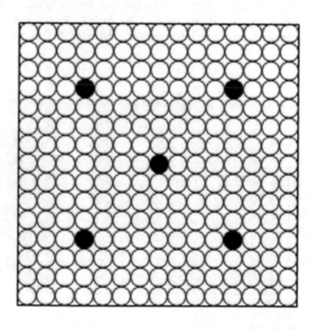

Step 1 Set the DE parameters ($F = 0.25$, $CR = 0.8$, see Sect. 5.2).

Step 2 Initialize the population of 5 individuals using the pattern that has been shown in Fig. 5.7 and the individual database array **T**, as an empty array.

Step 3 Compute the fitness values for each individual according to the fitness calculation strategy presented in Sect. 5.3. Since all individuals of the initial population fulfil rule 2 conditions, they are evaluated through a real fitness function (calculating the real SAD values).

Step 4 Update new evaluations in the individual database array **T**.

Step 5 Generate a new population of five individuals (trial population) considering the DE operators of mutation Eq. (5.2) and crossover Eq. (5.3).

Step 6 Compute fitness values for each individual by using the fitness calculation strategy presented in Sect. 5.3.

Step 7 Update new evaluations (rule 1 and rule 2) or estimations (rule 3) in the individual database array **T**.

Step 8 Select the fittest element between each individual and its corresponding trial counterpart according to Eq. (5.4) in order to obtain the final individual for next generation.

Step 9 If seven iterations have not been reached, then go back to Step 5; otherwise the best individual ($\hat{u}_{best}$, $\hat{v}_{best}$) from the final population is considered as the MV.

The presented DE-BM algorithm uses 40 different individuals (search locations) during the complete optimization process. However, only from 7 to 18 search locations are evaluated using the real fitness function (SAD evaluation) while the remaining positions are just estimated. Figure 5.8 shows two search-patterns examples that have been generated by the DE-BM approach. Such patterns exhibit

(a)   **(b)**

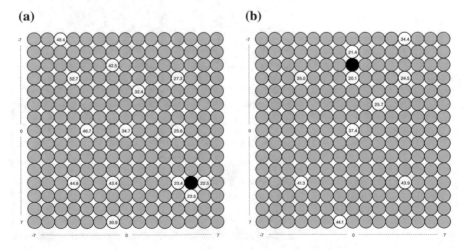

Fig. 5.8 Search-patterns generated by the DE-BM algorithm. **a** Pattern with solution $\hat{u}^1_{best} = 5$ and $\hat{v}^1_{best} = 4$. **b** Pattern with solution $\hat{u}^2_{best} = -5$ and $\hat{v}^2_{best} = 0$

the evaluated search-locations (rule 1 and 2) in white-cells, whereas the minimum location is marked in black. Grey-cells represent cells that have been estimated (rule 3) or not visited during the optimization process.

5.6   Experimental Results

This section presents the comparison between the presented DE-BM algorithm and other existing block matching algorithms. The simulations have been performed over the luminance component of popular video sequences that are listed in Table 5.1. Such sequences consist of different degrees and types of motion including QCIF (176 × 144), CIF (352 × 288) and SIF (352 × 240) respectively. The first four sequences are *Container, Carphone, Foreman* and *Akiyo* in QCIF format. The next two sequences are *Stefan* in CIF format and *Tennis* in SIF format. Among such sequences, *Container* has gentle, smooth and low motion changes and consists mainly of stationary and quasi-stationary blocks. *Carphone, Foreman* and *Akiyo* have moderately complex motion getting a "medium" category regarding its motion content. Rigorous motion which is based on camera panning with translation and complex motion content can be found in the sequences of *Stefan* and *Tennis*. Figure 5.9 shows a sample frame from each sequence.

Each picture frame is partitioned into macro-blocks with the sizes of 16×16 ($N = 16$) pixels for motion estimation, where the maximum displacement within the search range is ± 7 pixels in both horizontal and vertical directions.

Fig. 5.9   Test video sequences

In order to compare the performance of the DE-BM approach, different search algorithms such as FSA, TSS [10], 4SS [13], NTSS [11], BBGD [17], DS [14], NE [18], ND [20], LWG [24], GFSS [25] and PSO-BM [26] have been all implemented in our simulations. For comparison purposes, all six video sequences in Fig. 5.8 have been all used. All simulations are performed on a Pentium IV 3.2 GHz PC with 1 GB of memory.

In the comparison, two relevant performance indexes have been considered: the coding quality and the search efficiency.

5.6.1   Coding Quality

First, all algorithms are compared in terms of their coding quality which is characterized by the Peak-Signal-to-Noise-Ratio (PSNR) value. Such value indicates the reconstruction quality when motion vectors, which are computed through a BM approach, are used. In PSNR, the signal is the original data frames whereas the noise is the error introduced by the calculated motion vectors. The PSNR is thus defined as:

$$PSNR = 10 \cdot \log_{10}\left(\frac{255^2}{MSE}\right) \tag{5.8}$$

where MSE is the mean square between the original frames and those compensated by the motion vectors. Additionally, as an alternative performance index, it is used in the comparison the PSNR degradation ratio (D_{PSNR}). This ratio expresses in percentage (%) the level of mismatch between the PSNR of a BM approach and the PSNR of the FSA which is considered as reference. Thus the D_{PSNR} is defined as:

$$D_{PSNR} = -\left(\frac{PSNR_{FSA} - PSNR_{BM}}{PSNR_{FSA}}\right) \cdot 100\% \tag{5.9}$$

Table 5.2 shows the comparison of the PSNR values and the PSNR degradation ratios (D_{PSNR}) among several algorithms. The experiment considers six image sequences presented in Fig. 5.8. As it can be seen, in the case of the slow-moving sequence *Container*, the PSNR values (the D_{PSNR} ratios) of all BM algorithms are similar. For the medium motion content sequences such as *Carphone*, *Foreman* and *Akiyo*, the approaches consistent of fixed patterns (TSS, 4SS and NTSS) exhibit the worst PSNR value (high D_{PSNR} ratio) except for the DS algorithm. On the other hand, BM methods that use evolutionary algorithms (LWG, GFSS, PSO-BM and DE-BM) present the lowest D_{PSNR} ratio, only one step under the FSA approach which is considered as reference. Finally, approaches based on the error-function minimization (BBGD and NE) and pixel-decimation (ND), show an acceptable performance. For high motion sequences such as *Stefan* and *Tennis*, it is possible to draw same conclusions, as the medium-motion-content sequences can be observed. Since the motion content of these sequences is complex producing error surfaces with more than one minimum, the performance, in general, becomes worst for most of the algorithms. However, the PSNR values (or D_{PSNR} ratios) of the DS and the DE-BM approaches hold a better performance. As a summary of the coding quality performance, the last column of Table 5.2 presents the average PSNR degradation ratio (D_{PSNR}) obtained for all sequences. According to such values, the presented DE-BM method is superior to any other approach. Due to the computation complexity, the FSA is considered just as a reference.

Figure 5.10a–b shows the comparison of the frame-wise coding performance for the *Akiyo* and *Tennis* sequences (considering only 100 frames). Since the algorithms FSA, DE-BM, LWG, BBGD, NE and DS obtained good PSNR values in the coding performance analysis, they are exclusively considered at the comparison. According to these graphs, the coding quality of the DE-BM is better than the other algorithms (FSA, LWG, BBGD, NE and DS) whose PSNR values fluctuate heavily for some regions in the video sequences. The PSNR values of the DE-BM algorithm are lightly under the FSA approach which is used as reference. For a high motion sequence such as *Tennis*, we present a comparative result for frame-wise PSNR in Fig. 5.10b. Results exhibit a similar pattern as for the *Akiyo* sequence.

5.6.2 Search Efficiency

The search efficiency is used as a measurement of computational complexity at this chapter. The search efficiency is calculated by counting the average number of search points (or the average number of SAD computations) for the MV estimation. In Table 5.3, the search efficiency is compared. Just above FSA, some evolutionary algorithms such as LWG, GFSS and PSO-BM hold the highest number of search points per block. On the contrary, the presented DE-BM algorithm can be considered as a fast approach as it maintains a similar performance to BBGDS and DS. From data shown in Table 5.3, the average number of search locations, corresponding to the DE-BM method, represents the number of SAD evaluations (the number of SAD estimations are not considered whatsoever). Additionally, the last two columns of Table 5.3 present the number of search locations that have been averaged (over the six sequences) and their performance rank. According to these values, the presented DE-BM method is ranking in the second place, just below BBGDS. The average number of search points visited by the DE-BM algorithm ranges from 9.2 to 16.8, representing the 4 and the 7.4% respectively in comparison to the FSA method. Such results demonstrate that our approach can significantly reduce the number of search points. Hence, the DE-BM algorithm presented at this chapter is comparable to the fastest algorithms and delivers a similar precision to the FSA approach.

Table 5.2 PSNR values and D_{PSNR} comparison for all BM methods

Algorithm	Container		Carphone		Foreman		Akiyo		Stefan		Tennis		Total average (D_{PSNR})
	PSNR	D_{PSNR}	PSNR	D_{PSNR}	PNR	D_{PSNR}	PSNR	D_{PSNR}	PSNR	D_{PSNR}	PSNR	D_{PSNR}	
FSA	43.18	0	31.51	0	31.69	0	29.07	0	25.95	0	35.74	0	0
TSS	43.10	-0.20	30.27	-3.92	29.37	-7.32	26.21	-9.84	21.14	-18.52	30.58	-14.42	-9.03
4SS	43.12	-0.15	30.24	-4.01	29.34	-7.44	26.21	-9.84	21.41	-17.48	30.62	-14.32	-8.87
NTSS	43.12	-0.15	30.35	-3.67	30.56	-3.57	27.12	-6.71	22.52	-13.20	31.21	-12.65	-6.65
BBGD	43.14	-0.11	31.30	-0.67	31.00	-2.19	28.10	-3.33	25.17	-3.01	33.17	-7.17	-2.74
DS	43.13	-0.13	31.26	-0.79	31.19	-1.59	28.00	-3.70	24.98	-3.73	33.98	-4.92	-2.47
NE	43.15	-0.08	31.36	-0.47	31.23	-1.47	28.23	-2.89	25.22	-2.81	33.88	-5.19	-2.15
ND	43.15	-0.08	31.35	-0.50	31.20	-1.54	28.21	-2.96	25.21	-2.86	33.79	-5.43	-2.22
LWG	43.16	-0.06	31.40	-0.36	31.31	-1.21	28.55	-1.80	25.41	-2.09	33.97	-4.95	-1.74
GFSS	43.15	-0.06	31.38	-0.40	31.29	-1.26	28.32	-2.58	25.34	-2.36	33.87	-5.23	-1.98
PSO-BM	43.15	-0.07	31.39	-0.38	31.27	-1.34	28.33	2.55	25.39	-2.15	33.91	-5.11	-1.93
DE-BM	43.17	-0.04	31.47	-0.13	31.51	-0.58	28.98	-1.00	25.85	-0.78	34.77	-4.27	-1.13

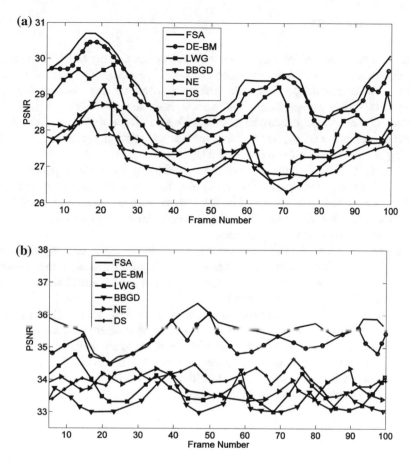

Fig. 5.10 Frame-wise performance comparison between different BMAs on sequence: **a** Akiyo and **b** Tennis, considering only 100 frames

Table 5.3 Averaged number of visited search points per block for all ten BM methods

Algorithm	Container	Carphone	Foreman	Akiyo	Stefan	Tennis	Total average	Rank
FSA	225	225	225	225	225	225	225	12
TSS	25	25	25	25	25	25	25	7
4SS	19	25.5	24.8	27.3	29.3	31.5	26.3	8
NTSS	17.2	21.8	22.1	23.5	25.4	26.1	22.6	6
BBGD	8.1	11.5	12.5	10.2	15.2	17.1	12.43	1
DS	7.5	12.5	13.4	11.8	16.2	17.5	13.15	3
NE	11.7	13.8	14.2	14.5	19.2	20.2	15.6	5
ND	10.8	13.4	13.8	14.1	18.4	19.1	14.9	4
LWG	75	75	75	75	75	75	75	11
GFSS	60	60	60	60	60	60	60	10
PSO-BM	32.5	48.5	48.1	48.5	52.2	52.2	47	9
DE-BM	9.2	12.2	12.2	12.5	16.1	16.8	13.14	2

5.7 Conclusions

In this chapter, a new algorithm based on Artificial Bee Colony (ABC) is presented to reduce the number of search locations in the BM process. The algorithm uses a simple fitness calculation approach which is based on the Nearest Neighbor Interpolation (NNI) algorithm. The method estimates the fitness value (SAD operation) for several candidate solutions (search locations). As a result, the approach can substantially reduce the number of function evaluations still preserving the good search capabilities of ABC.

Since the presented algorithm does not consider any fixed search pattern or any other assumption, a high probability for finding the true minimum (accurate motion vector) is expected with no regard of the movement complexity contained in the sequence. Therefore, the chance of being trapped into a local minimum is reduced in comparison to other BM algorithms.

The performance of DE-BM has been compared to other existing BM algorithms. Experimental results demonstrate the high performance of the presented method in finding accurate motion vectors (distortion performance) yet demanding competitive time rates (search efficiency).

Although the experimental results indicate that the ABC-BM method can yield better results on complicated sequences, it should be noticed that the aim of our chapter is not intended to beat all the BM methods which have been presented earlier, but to show that the fitness approximation can effectively serve as an attractive alternative to evolutionary algorithms for solving complex optimization problems demanding few function evaluations.

References

1. Dimitrios Tzovaras, Ioannis Kompatsiaris, Michael G. Strintzis. 3D object articulation and motion estimation in model-based stereoscopic videoconference image sequence analysis and coding. Signal Processing: Image Communication, 14(10), 1999, 817–840.
2. Barron, J.L., Fleet, D.J., Beauchemin, S.S., 1994. Performance of optical flow techniques. Int. J. Comput. Vision 12 (1), 43–77.
3. J. Skowronski. Pel recursive motion estimation and compensation in subbands. Signal Processing: Image Communication 14, (1999), 389–396.
4. MPEG1, Information Technology Coding of Moving Pictures and Associated Audio for Digital Storage Media At Up To About 1.5 mbit/s—Part 2: Video, JTC1/SC29/WG11, ISO/IEC11172-2 (MPEG-1 Video),1993.
5. MPEG2, Generic Coding of Moving Pictures and Associated Audio Information—Part 2: Video, ITU-T and ISO/IECJTC1, ITURec. H.262—ISO/IEC 13818-2(MPEG-2Video), 1994.
6. MPEG4, Information Technology Coding of Audio Visual Objects Part 2: Visual, JTC1/SC29/WG11, ISO/IEC14469-2(MPEG-4Visual), 2000.
7. H261,Video Codec for Audio visual Services at px64 kbit/s, ITU-T SG15, ITU-TRec.H.261, seconded, 1993.
8. I.-T.R., H.263, Video Coding for Low Bit Rate Communication, ITU-T SG16, ITU-TRec. H.263, thirded, 2000.

9. J. R. Jain and A. K. Jain, Displacement measurement and its application in interframe image coding, IEEE Trans. Commun., vol. COM-29, pp. 1799–1808, Dec. 1981.

10. H.-M. Jong, L.-G. Chen, and T.-D. Chiueh, "Accuracy improvement and cost reduction of 3-step search block matching algorithm for video coding," IEEE Trans. Circuits Syst. Video Technol., vol. 4, pp. 88–90, Feb. 1994.

11. Renxiang Li, Bing Zeng, and Ming L. Liou, "A New Three-Step Search Algorithm for Block Motion Estimation", IEEE Trans. Circuits And Systems For Video Technology, vol 4., no. 4, pp. 438–442, August 1994.

12. Jianhua Lu, and Ming L. Liou, "A Simple and Efficient Search Algorithm for Block-Matching Motion Estimation", IEEE Trans. Circuits And Systems For Video Technology, vol 7, no. 2, pp. 429–433, April 1997.

13. Lai-Man Po, and Wing-Chung Ma, "A Novel Four-Step Search Algorithm for Fast Block Motion Estimation", IEEE Trans. Circuits And Systems For Video Technology, vol 6, no. 3, pp. 313–317, June 1996.

14. Shan Zhu, and Kai-Kuang Ma, " A New Diamond Search Algorithm for Fast Block-Matching Motion Estimation", IEEE Trans. Image Processing, vol 9, no. 2, pp. 287–290, February 2000.

15. Yao Nie, and Kai-Kuang Ma, Adaptive Rood Pattern Search for Fast Block-Matching Motion Estimation, IEEE Trans. Image Processing, vol 11, no. 12, pp. 1442–1448, December 2002.

16. Yi-Ching L., Jim L., Zuu-Chang H. Fast block matching using prediction and rejection criteria. Signal Processing, 89, (2009), pp 1115–1120.

17. Liu, L., Feig, E. A block-based gradient descent search algorithm for block motion estimation in video coding, IEEE Trans. Circuits Syst. Video Technol., 6(4),(1996),419–422.

18. Salia, A., Mukherjee, J., Sural, S. A neighborhood elimination approach for block matching in motion estimation, Signal Process Image Commun, (2011), 26, 8–9, 2011, 438–454.

19. K.H.K. Chow, M.L. Liou, Generic motion search algorithm for video compression, IEEE Trans. Circuits Syst. Video Technol. 3, (1993), 440–445.

20. A. Saha, J. Mukherjee, S. Sural. New pixel-decimation patterns for block matching in motion estimation. Signal Processing: Image Communication 23 (2008)725–738.

21. Y. Song, T. Ikenaga, S. Goto. Lossy Strict Multilevel Successive Elimination Algorithm for Fast Motion Estimation. IEICE Trans. Fundamentals E90(4), 2007, 764–770.

22. J.H. Holland, Adaptation in Natural and Artificial Systems, University of Michigan Press, Ann Arbor, MI, 1975.

23. J. Kennedy, R.C. Eberhart, Particle swarm optimization, in: Proceedings of the 1995 IEEE International Conference on Neural Networks, vol. 4, 1995, pp. 1942–1948.

24. Chun-Hung, L., Ja-Ling W. A Lightweight Genetic Block-Matching Algorithm for Video Coding. IEEE Transactions on Circuits and Systems for Video Technology, 8(4), (1998), 386–392.

25. Wu, A., So, S. VLSI Implementation of Genetic Four-Step Search for Block Matching Algorithm. IEEE Transactions on Consumer Electronics, 49(4), (2003), 1474–1481.

26. Yuan, X., Shen, X. Block Matching Algorithm Based on Particle Swarm Optimization. International Conference on Embedded Software and Systems (ICESS2008), 2008, Sichuan, China.

27. Storn R, Price K. Differential evolution – a simple and efficient adaptive scheme for global optimization over continuous spaces. Technical Rep. No. TR-95-012, International Computer Science Institute, Berkley (CA). (1995).

28. Babu B, Munawar S. Differential evolution strategies for optimal design of shell-and-tube heat exchangers. Chem Eng Sci. 62(14):3720–39. (2007).

29. Mayer D, Kinghorn B, Archer A. Differential evolution – an easy and efficient evolutionary algorithm for model optimization. Agr Syst, 83:315–28. (2005).

30. Kannan S, Mary Raja Slochanal S, Padhy N. Application and comparison of metaheuristic techniques to generation expansion planning problem. IEEE Trans Power Syst, 20(1):466–75. (2003).

31. Chiou J, Chang C, Su C. Variable scaling hybrid differential evolution for solving network reconfiguration of distribution systems. IEEE Trans Power Syst, 20(2):668–74. (2005).
32. Chiou J, Chang C, Su C. Ant direct hybrid differential evolution for solving large capacitor placement problems. IEEE Trans Power Syst, 19(4):1794–800. (2004).
33. Ursem R, Vadstrup P. Parameter identification of induction motors using differential evolution. In: Proceedings of the 2003 congress on evolutionary computation (CEC'03), vol. 2, Canberra, Australia, p. 790–6. (2003).
34. Babu B, Angira R, Chakole G, Syed Mubeen J. Optimal design of gas transmission network using differential evolution. In: Proceedings of the second international conference on computational intelligence, robotics, and autonomous systems (CIRAS-2003), Singapore. (2003).
35. Zelinka, I., Chen, G., Celikovsky, S.: Chaos synthesis by means of evolutionary algorithms. Int. J. Bifurcat Chaos 4, 911–942 (2008).
36. E. Cuevas, D. Zaldivar, M. Pérez-Cisneros. A novel multi-threshold segmentation approach based on differential evolution optimization. Expert Systems with Applications 37 (2010) 5265–5271.
37. Jin, Y. Comprehensive survey of fitness approximation in evolutionary computation. Soft Computing, 9, (2005), 3–12.
38. Yaochu Jin. Surrogate-assisted evolutionary computation: Recent advances and future challenges. Swarm and Evolutionary Computation, 1, (2011), 61–70.
39. J. Branke, C. Schmidt. Faster convergence by means of fitness estimation. Soft Computing 9, (2005), 13–20.
40. Zhou, Z., Ong, Y., Nguyen, M., Lim, D. A Study on Polynomial Regression and Gaussian Process Global Surrogate Model in Hierarchical Surrogate-Assisted Evolutionary Algorithm, IEEE Congress on Evolutionary Computation (ECiDUE'05), Edinburgh, United Kingdom, September 2–5, 2005.
41. Ratle, A. Kriging as a surrogate fitness landscape in evolutionary optimization. Artificial Intelligence for Engineering Design, Analysis and Manufacturing, 15, (2001), 37–49.
42. Lim, D., Jin, Y., Ong, Y., Sendhoff, B. Generalizing Surrogate-assisted Evolutionary Computation, IEEE Transactions on Evolutionary Computation, 14(3), (2010), 329–355.
43. Ong, Y., Lum, K., Nair, P. Evolutionary Algorithm with Hermite Radial Basis Function Interpolants for Computationally Expensive Adjoint Solvers, Computational Optimization and Applications, 39(1), (2008), 97–119.
44. Luoa, C., Shao-Liang, Z., Wanga, C., Jiang, Z. A metamodel-assisted evolutionary algorithm for expensive optimization. Journal of Computational and Applied Mathematics, doi:10.1016/j.cam.2011.05.047, (2011).
45. Goldberg, D. E. Genetic algorithms in search, optimization and machine learning. Menlo Park, (1989) CA: Addison-Wesley Professional.
46. Li, X., Xiao, N., b, Claramunt, C., Lin, H. Initialization strategies to enhancing the performance of genetic algorithms for the p-median problem, Computers & Industrial Engineering, (2011), doi:10.1016/j.cie.2011.06.015.
47. Xiao, N. A unified conceptual framework for geographical optimization using evolutionary algorithms. Annals of the Association of American Geographers, 98, (2008), 795–817. doi:10.1007/s10732-008-9080-4.

Chapter 6
Photovoltaic Cell Design

Abstract In order to improve the performance of solar energy systems, accurate modeling of current versus voltage (I–V) characteristics of solar cells has attracted the attention of various researches. The main drawback in accurate modeling is the lack of information about the precise parameter values which indeed characterize the solar cell. Since such parameters cannot be extracted from the datasheet specifications, an optimization technique is necessary to adjust experimental data to the solar cell model. Considering the I–V characteristics of solar cells, the optimization task involves the solution of complex non-linear and multi-modal objective functions. Several optimization approaches have been presented to identify the parameters of solar cells. However, most of them obtain sub-optimal solutions due to their premature convergence and their difficulty to overcome local minima in multi-modal problems. This chapter describes the use of the Artificial Bee Colony (ABC) algorithm to accurately identify the solar cells' parameters. The ABC algorithm is an evolutionary method inspired by the intelligent foraging behavior of honeybees. In comparison with other evolutionary algorithms, ABC exhibits a better search capacity to face multi-modal objective functions. In order to illustrate the proficiency of the presented approach, it is compared to other well-known optimization methods. Experimental results demonstrate the high performance of the presented method in terms of robustness and accuracy.

6.1 Introduction

The increase in the cost of fossil fuels and their probable depletion, air pollution, global warming phenomenon, and severe environmental laws have resulted in renewable energy sources gaining the attention of many nations to produce electricity. Solar energy is one of the most promising renewable sources that is currently being used worldwide to contribute to meeting rising demands for electric power. It has been reported that solar photovoltaic (PV) is the fastest growing power-generation

© Springer International Publishing AG 2017
E. Cuevas et al., *Evolutionary Computation Techniques:*
A Comparative Perspective, Studies in Computational Intelligence 686,
DOI 10.1007/978-3-319-51109-2_6

technology in the world, with an annual average increase of 50% between 2004 and 2011 [1]. PV is not only capable of directly converting solar energy to electricity but also is an emission-free distributed generation unit that would supply power at the load site.

Solar cell accurate modeling has received significant attention in recent years [2–6]. The modeling of PV cells consists in two steps: the mathematical model formulation and the accurate estimation of their parameter values. For the mathematical model, the Current versus Voltage (I–V) characteristics that rule the behavior of a solar cell is considered. Several approaches have been presented in order to model such a behavior from different point of views [7–12].

In practical terms, there exist two equivalent electronic circuits that model the behavior of a solar cell. Such circuits are known as single diode (SD) and double diode (DD) models [13]. Irrespective of the model selected, it is necessary to estimate or identify all its parameters such as photo-generated current, diode saturation current, series resistance, and diode ideality factor. Depending on the model (SD or DD), two different sets of parameters must be identified: five for the SD and seven for the DD. The main problem is to identify the optimal parameter values which, when applied to the selected model, produce the best possible approximation to the experimental data obtained by the true solar cell [13].

The methods employed to solve the problem of PV parameter identification can be divided in two groups: deterministic and heuristic. Some examples of deterministic methods involve methods such as least squares [14], Lambert W-functions [15], and the iterative curve fitting [16]. Deterministic techniques impose several model restrictions such as convexity and differentiability in order to be correctly applied. Therefore, they are very sensitive to the initial solution, and most often lead to local optima. As an alternative to deterministic-based techniques, the problem of PV parameter identification has also been handled through heuristic methods. In general, they have demonstrated that they deliver better results than those based on deterministic approaches considering accuracy and robustness [13, 17–21]. In the literature, several heuristic approaches have been presented in order to solve the problem of solar cell parameter identification. Such methods include genetic algorithms (GA) [17, 18], particle swarm optimization (PSO) [19, 20], simulated annealing (SA) [21], and Harmony Search (HS) [13]. Although heuristic methods present a higher probability of obtaining a global solution in comparison with deterministic ones, they have important limits [18]. In case of GA and PSO, they maintain a trend that concentrates toward local optima, since their elitist mechanism forces premature convergence [22, 23]. Such a behavior becomes worse when the optimization algorithm faces multi-modal functions [24, 25]. On the other hand, due to the fact that SA and HS are single-searcher algorithms, their performance is sensitive to the starting point of the search, having a lower probability to localize the global minimum in multi-modal problems than population algorithms such as GA and PSO [26, 27]. Therefore, GA, PSO, SA, and HS present a bad performance when they are applied to multi-modal and noisy objective functions.

In order to identify the PV parameters as an optimization problem, it is necessary to define an objective function. Such an objective function is built by using experimental data extracted from I–V measurements of the solar cell. Since experimental data contain noise as a consequence of an imperfect data collection process, the objective function obtained presents high multi-modal and noisy characteristics [28, 29]. Under these circumstances, most of the heuristic approaches present a bad performance [30].

In this chapter, an alternative approach using the Artificial Bee Colony (ABC) [31] method for determining the parameters of a solar cell is presented. The ABC is an evolutionary algorithm inspired by the intelligent behavior of honey bees. The performance of the ABC has been compared to other evolutionary methods such as GA and PSO [32, 33]. The results have shown that ABC produces optimal solutions when it faces multi-modal and noisy optimization problems. Such characteristics have motivated the use of ABC to solve different types of engineering problems within several fields [34–39]. One relevant advantage of the ABC method is that it does not follow a local strategy for computing new solutions. Instead, the ABC method uses a set of operators to build solutions from random operations avoiding falling into local optimal.

ABC consists of three essential components: food source positions, nectar amount, and several honey-bee classes. Each food source position represents a feasible solution for the problem under consideration. The nectar amount for a food source represents the quality of such a solution (represented by a fitness value). Each bee class symbolizes one particular operation for generating new candidate food source positions (i.e., candidate solutions). The ABC algorithm starts by producing a randomly distributed initial population (food source locations). After initialization, an objective function evaluates whether such candidates represent an acceptable solution (nectar amount) or not. Guided by the values of such an objective function, candidate solutions are evolved through different ABC operations (honey-bee types) until a termination criterion is met.

This chapter presents the use of ABC to accurately estimate the parameter of solar Cells. In the approach, the estimation process is considered as an optimization problem. The presented approach encodes the parameters of the solar cell as a candidate solution. An objective function evaluates the matching quality between a candidate solution and the experimental data. Guided by the values of this objective function, the set of encoded candidate solutions is evolved by using the operators defined by ABC so that the parameters that produce the best possible approximation to the I–V measurements obtained by the true solar cell can be found. In order to illustrate the proficiency of the presented approach, it is compared to other well-known optimization methods. Experimental evidence shows that ABC exhibits no sensitivity to noisy conditions and high performance in terms of robustness and accuracy.

The remainder of the chapter is organized as follows. In Sect. 6.2, the problem of solar cell identification is defined. Section 6.3 describes the ABC algorithm. In Sect. 6.4, the problem of solar cell identification is translated to an optimization task. Section 6.5 presents the experimental results and comparisons. Finally, in Sect. 6.6, the conclusions are stated.

6.2 Solar Cell Modeling

The modeling of PV cells consists in two steps: the mathematical model formulation and the accurate estimation of their parameter values. For the mathematical model, it is considered the Current versus Voltage (*I–V*) characteristics that rule the behavior of a solar cell. Several approaches have been presented in order to model such a behavior under different point of views. In practical terms, there exist two models: single diode (SD) and double diode (DD) [13]. In this section these models are described and their objective functions are formulated.

6.2.1 Double Diode Model

Ideally, a solar cell is modeled considering a photo-generated (I_{ph}) current source which is shunted with a rectifying diode. However, in practice, the modeling of SC needs to consider detailed aspects involved in the conversion process. For practical terms, the current source I_{ph} is shunted by another diode which is used to model the space charge recombination current and a shunt leakage resistor to account for the partial short circuit current path near the cell's edges due to the semiconductor

Fig. 6.1 Double diode model of solar cells

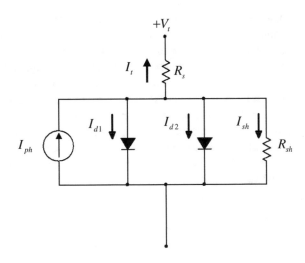

impurities and non-idealities. The structure of solar cells includes metal contacts which combined with semiconductor materials increase the resistance. Such a resistance is represented in the model by a resistor connected in series with the cell shunt elements [31]. Figure 6.1 shows the equivalent circuit for the DD model.

According to Fig. 6.1, the cell terminal current is computed as follows:

$$I_t = I_{ph} - I_{d1} - I_{d2} - I_{sh}, \tag{6.1}$$

where I_t is the terminal current, I_{ph} the photo-generated current, I_{d1}, I_{d2} is the first and second diode currents whereas I_{sh} is the shunt resistor current. In order to appropriately model the solar cell, there is used the Shockley diode equation; hence, Eq. 6.1 is rewritten as it is shown in Eq. 6.2.

$$I_t = I_{ph} - I_{sd1} \left[\exp\left(\frac{q(V_t + R_s \cdot I_t)}{n_1 \cdot k \cdot T} \right) - 1 \right]$$
$$\dots - I_{sd2} \left[\exp\left(\frac{q(V_t + R_s \cdot I_t)}{n_2 \cdot k \cdot T} \right) - 1 \right] - \frac{V_t + R_s \cdot I_t}{R_{sh}}, \tag{6.2}$$

where I_{sd1} and I_{sd2} are the diffusion and saturation current, respectively. V_t is the terminal voltage whereas the series and shunt resistances are represented by R_s and R_{sh} respectively. According to the Shockley diode equation, $q = 1.602 \times 10^{-19}$ (coulombs) is the magnitude of charge on an electron, $k = 1.380 \times 10^{-23}$ (J/K) is the Boltzmann constant, n_1 and n_2 are the diffusion and recombination diode ideality factors, respectively. Finally, T is the cell temperature (K). Therefore, Eq. 6.2 has seven unknown parameters (R_s, R_{sh}, I_{ph}, I_{sd1}, I_{sd2}, n_1, and n_2). An accurate identification of such parameters allows projecting the optimal performance of a solar cell, for that reason the estimation process is an important task.

6.2.2 Single Diode Model

In a solar cell, the diffusion (I_{sd1}) and saturation (I_{sd2}) currents are different and independent. However, in order to formulate a condensed model, both currents are combined by using a non-physical ideality factor n [13, 17, 21]. Such reduced formulation, shown in Fig. 6.2, is known as the single diode (SD) model which is widely used for modeling solar cells due to its simplicity. Different to the DD, the SD model has only five parameters to be identified.

Under the SD model, Eq. 6.2 is reduced to the following equation:

$$I_t = I_{ph} - I_{sd} \left[\exp\left(\frac{q(V_t + R_s \cdot I_t)}{n \cdot k \cdot T} \right) - 1 \right] - \frac{V_t + R_s \cdot I_t}{R_{sh}} \tag{6.3}$$

Consequently, the parameters to be identified are R_s, R_{sh}, I_{ph}, I_{sd}, and n.

Fig. 6.2 Single diode model
of solar cells

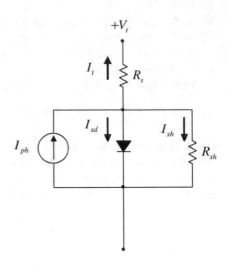

6.2.3   Parameter Identification of a Solar Cell
as an Optimization Problem

The problem of modeling solar cells consists in accurately identifying the parameters of Eqs. 6.2 and 6.3. In the presented approach, the problem of parameter identification is considered as an optimization problem where it is sought the parameter set that produces the best approximation to the I–V measurements obtained by the true solar cell. Therefore, it is necessary to define an objective function that evaluates the matching quality between a candidate parameter set and the experimental data. In this chapter, the problem of solar cell modeling is approached considering the SD as well as the DD model. Thus, Eqs. 6.2 and 6.3 must be rewritten in order to reflex their deviation with regard to experimental data. Thereby, for the DD model, the deviation function is defined as follows:

$$f_{DD}(V_t, I_t, \mathbf{x}) = I_t - I_{ph} + I_{sd1} \left[\exp \left(\frac{q(V_t + R_s \cdot I_t)}{n_1 \cdot k \cdot T} \right) - 1 \right]$$
$$\ldots + I_{sd2} \left[\exp \left(\frac{q(V_t + R_s \cdot I_t)}{n_2 \cdot k \cdot T} \right) - 1 \right] + \frac{V_t + R_s \cdot I_t}{R_{sh}}, \tag{6.4}$$

whereas for the SD model such function is formulated as Eq. 6.5.

$$f_{SD}(V_t, I_t, \mathbf{x}) = I_t - I_{ph} + I_{sd} \left[\exp \left(\frac{q(V_t + R_s \cdot I_t)}{n \cdot k \cdot T} \right) - 1 \right] + \frac{V_t + R_s \cdot I_t}{R_{sh}} \tag{6.5}$$

In both functions (f_{DD} and f_{SD}), the values of V_t and I_t are experimentally collected from the solar cell. Therefore, the parameter estimation is a process that minimizes the difference between the measured data and the calculated current by adjusting the model parameters. Considering that the number of experimental data is N, the objective function can be formulated by the Root Mean Square Error (*RMSE*) as:

$$RMSE(\mathbf{x}) = \sqrt{\frac{1}{N}\sum_{i=1}^{N}\left(f_M^i\left(V_t^i, I_t^i, \mathbf{x}\right)\right)},\qquad(6.6)$$

where M is the model type DD or SD, $\mathbf{x} = \left[R_s, R_{sh}, I_{ph}, I_{sd1}, I_{sd2}, n_1, n_2\right]$ is the model parameters for DD and $\mathbf{x} = \left[R_s, R_{sh}, I_{ph}, I_{sd}, n\right]$ for SD.

As it is formulated in Eq. 6.6, the objective function is built by using experimental data, extracted from *I–V* measurements of the solar cell. Since experimental data contain noise as a consequence of an imperfect data collection process, the objective function obtained presents high multi-modal and noisy characteristics [28, 29]. Under these circumstances, most of the heuristic approaches present a bad performance [30].

6.3   Artificial Bee Colony Algorithm

In this chapter, an alternative approach that uses the Artificial Bee Colony (ABC) [31] method for determining the parameters of a Solar Cell is introduced. The ABC is an evolutionary algorithm inspired by the intelligent behavior of honey-bees. The ABC algorithm has demonstrated to produce optimal solutions when it faces multi-modal and noisy optimization problems.

The ABC algorithm assumes the existence of a set of operations that may resemble some features of the honeybee behavior. For instance, each solution within the search space includes a parameter set representing food source locations. The "fitness value" refers to the food source quality that is linked to the food's location. The process mimics the bee's search for valuable food sources yielding an analogous process for finding the optimal solution.

6.3.1   Biological Profile

The minimal model for a honeybee colony consists of three classes: employed bees, onlooker bees and scout bees. The employed bees will be responsible for investigating the food sources and sharing the information with recruit onlooker bees. They, in turn, will make a decision on choosing food sources by considering such information. The food source having a higher quality will have a larger chance to be selected by onlooker bees than those showing a lower quality. An employed bee, whose food source is rejected as low quality by employed and onlooker bees, will change to a scout bee to randomly search for new food sources. Therefore, the exploitation is driven by employed and onlooker bees while the exploration is maintained by scout bees.

6.3.2   Description of the ABC Algorithm

Similar to other swarm-based approaches, the ABC algorithm is an iterative process. It starts with a population of randomly generated solutions or food sources. The following three operations are applied until a termination criterion is met [40]:

1. Send the employed bees.
2. Select the food sources using the onlooker bees.
3. Determine the scout bees.

6.3.2.1 Initializing the Population

The first step of the algorithm is to initialize the population of N_p food sources. Every food source is a d-dimensional vector containing the parameters values to be optimized. Such values are randomly and uniformly distributed between a bounded space.

$$
x_{i,j} = l_j + \text{rand}(0, 1) \cdot (u_j - l_j),
$$
$$
j = 1, 2, \ldots, d; \quad i = 1, 2, \ldots, N_p,
\tag{6.7}
$$

where the index i corresponds to ith food source and j is the jth dimension of the search space. l_j and u_j are the lower and the upper bound in each dimension.

6.3.2.2 Send Employed Bees

The employed bees are used to generate new solutions; the number of this kind of bees is equal to the number of food sources. According with the literature [31, 32] the entire population is divided in two $(N_p/2)$, one part corresponds to the employed bees and the rest to the onlooker bees.

$$
v_{i,j} = x_{i,j} + \phi_{i,j}(x_{i,j} - x_{k,j}), \quad \forall i \neq k
$$
$$
k \in \text{rand}\{1, N_p\}, \ j \in \{1, 2, \ldots, d\}
\tag{6.8}
$$

The parameter $\phi_{j,i}$ is a random value selected between $[-1, 1]$, then to generate the new source food using the employed bee operator, in a randomly way is selected a k food source in the j dimension. If a parameter of $v_{i,j}$ exceeds the boundaries, it should be adjusted in order to fit the appropriate range. After this process is calculated the fitness value associated with each solution. For a minimization problem it can be obtained using the following expression:

$$fit_i = \begin{cases} \dfrac{1}{1+J_i} & \text{if } J_i \geq 0 \\ 1 + \text{abs}(J_i) & \text{if } J_i < 0 \end{cases} \tag{6.9}$$

where J_i is the objective function to be minimized. The next process consist in apply a greedy selection between the values v_i and x_i, that means: if the nectar-amount (fitness) of v_i is better, then the solution x_i is replaced by v_i otherwise, x_i is preserved.

6.3.2.3   Select the Food Sources by the Onlooker Bees

In order to describe the onlooker phase, first it is necessary to explain that the number of onlooker bees corresponds to the food source number. In this way the food sources are modified several times depending on the fitness value (Eq. 6.9). For a food source could be selected, it is necessary to obtain a probability factor that is computed based on the fitness.

$$fit_i = \begin{cases} \dfrac{1}{1+J_i} & \text{if } J_i \geq 0 \\ 1 + \text{abs}(J_i) & \text{if } J_i < 0 \end{cases} \tag{6.10}$$

Here, fit_i corresponds to the fitness value of the i th food source and is related to the objective function of the food source i. If the fitness of a food source increases, then the probability of be selected by an onlooker is bigger. When a food source is selected a new value is obtained using Eq. 6.2, its fitness is computed and the greedy process is applied to modify (or not) its position.

6.3.2.4   Determine Scout Bees

The final step is the scout bee process. Here the bees are applied if a food source i cannot be improved through a predetermined trial "limit" number, then the food source is considered to be abandoned and instead to be modified by and onlooker bee, is modified by an scout bee using Eq. 6.1. The predefined "*limit*" is a counter assigned to each food source and is incremented when the fitness is not improved.

6.3.3   ABC Computational Procedure

The complete ABC Algorithm can be summarized by the instructions listed in Algorithm 6.1.

Algorithm 6.1 Artificial bee colony method

1:	Generate the initial population $x_{i,j}$, $j = 1,2,...,d$; $i = 1,2,...,N_p$, Eq. 6.7.
2:	Evaluate the initial population with regard to the objective function
3:	Set cycle to 1
4:	**repeat**
5:	**for** each employed bee
	{
	Produce new solution v_i by using Eq. 6.8
	Calculate the value fit_i (Eq. 9)
	Apply greedy selection process
	}
6:	Calculate the probability values $Prob_i$ for the produced solutions by Eq. 10
7:	**for** each onlooker bee
	{
	Select a solution i depending on $Prob_i$
	Produce new solution v_i
	Calculate the value fit_i
	Apply greedy selection process
	}
8:	**if** (there is an abandoned solution) **then** replace it with a new random solution produced by a scout using Eq. 6.7
9:	Memorize the best solution so far
10:	cycle = cycle + 1
11:	**until** cycle = maximum iteration number

6.4 Parameter Identification of Solar Cells Using ABC

6.4.1 Problem Statement

The presented approach encodes the parameters of the solar cell as a candidate solution. The representation of such a candidate solution (food source) depends on the model type: DD or SD. Therefore, each food source uses seven elements for the DD formulation and five for the SD model, as decision variables within the optimization algorithm. Thus, the estimation task is faced as an optimization problem which can be stated as follows:

$$
\begin{aligned}
\text{minimize:} \quad & RMSE(\mathbf{X}), \; \mathbf{X} = [\mathbf{x}_1, \mathbf{x}_2, \ldots, \mathbf{x}_{Np}], \quad \mathbf{x}_{wi} = [x_{i,1}, x_{i,2}, \ldots x_{i,d}] \; d \in [5, 7], \\
\text{subject to:} \quad & d = 5 \; (\text{SD}) \qquad\qquad\qquad\qquad\quad d = 7 \; (\text{DD}) \\
& 0 \le x_{i,1}(R_s) \le 0.5 \qquad\qquad\quad\; 0 \le x_{i,1}(R_s) \le 0.5 \\
& 0 \le x_{i,2}(R_{sh}) \le 100 \qquad\qquad\; 0 \le x_{i,2}(R_{sh}) \le 100 \\
& 0 \le x_{i,3}(I_{ph}) \le 1 \qquad\qquad\quad\;\; 0 \le x_{i,3}(I_{ph}) \le 1 \\
& 0 \le x_{i,4}(I_{sd}) \le 1 \qquad\qquad\quad\;\; 0 \le x_{i,4}(I_{sd1}) \le 1 \\
& 1 \le x_{i,5}(n) \le 2 \qquad\qquad\qquad 0 \le x_{i,5}(I_{sd2}) \le 1 \\
& \qquad\qquad\qquad\qquad\qquad\qquad\;\; 1 \le x_{i,6}(n_1) \le 2 \\
& \qquad\qquad\qquad\qquad\qquad\qquad\;\; 1 \le x_{i,7}(n_1) \le 2
\end{aligned}
$$

$$(6.11)$$

where N_p and d are the population size and the number of dimensions, respectively.

6.4.2 Computational Approach

The presented algorithm has been implemented considering the two different solar cell models (SD and DD) whereas its efficiency is evaluated using de RMSE criterion. As optimization technique, the ABC method is used to solve the problem of parameter identification defined by Eq. 6.11. The computational procedure of the presented approach can be summarized into the Algorithm 6.2.

Algorithm 6.2 Computational approach

1:	Read the N experimental data values of V_t and I_t. Then, Store them in the vector $\mathbf{ED} = [ED_1, ED_2, \dots ED_N]$, $ED_i = [V_t^i, I_t^i]$.
2:	Initialize the ABC parameters: $c_{\max}$ (maximum iteration number), *limit* (limit to declare an abandoned solution), d (number of dimensions) and N_p (population size).
3:	Initialize the population $\mathbf{X}$ of N_p random candidate solutions with d dimensions depending on the solar cell model ($d=5$ for SD and $d=7$ for DD).
4:	Evaluate the initial population with regard to the objective function
5:	Set cycle to 1
6:	**repeat**
7:	**for** each employed bee
	{
	Produce new solution v_i by using Eq. 6.8
	Calculate the value fit_i (Eq. 6.9)
	Apply greedy selection process
	}
8:	Calculate the probability values $Prob_i$ for the produced solutions by Eq. 6.10);
9:	**for** each onlooker bee
	{
	Select a solution i depending on $Prob_i$
	Produce new solution v_i
	Calculate the value fit_i
	Apply greedy selection process
	}
10:	**if** (a candidate solution does not change in more than *limit* iterations) **then** replace it with a new random solution produced by a scout using Eq. 6.7
11:	Memorize the best solution so far
12:	cycle = cycle + 1
13:	**until** cycle = $c_{\max}$

6.5 Experimental Results

In order to prove the performance of the presented approach, the algorithm was tested using a commercial silicon solar cell (RTC France), with a diameter of 57 mm. During the data collection process, it is considered that the solar cell operates under the following operating conditions: 1 sun $(1000\,\text{W}/\text{m}^2)$ at $T = 33\,^\circ\text{C}$. In this section, three experiments are conducted. In the first one, the presented approach is employed to extract the cell parameters using the single and double diode models. Then, the results of the ABC-based approach are compared with other well known optimization algorithms. Conclusions of the experimental comparison are validated through a statistical test that supports the truthfulness of the results. Finally, in the third experiment, the robustness of the presented approach is tested under different noise conditions. In order to conduct such experiments, the ABC is configured considering the parameter values shown in Table 6.1. Once they have been determined after intensive tests, they are kept for all experiments.

The parameter *limit* is computed as $N_p \cdot d$, where N_p and d are the population size and the number of dimensions, respectively. In the experiments, the stop criterion is the maximum iteration number $c_{\max}$. However, if the fitness value for the best candidate solution remains unspoiled in 10% of the total number of $c_{\max}$, then the algorithm is stopped.

6.5.1 ABC Experimental Results

This experiment presents the results of the presented approach when it is employed to extract the cell parameters considering the single and double diode models. To this end, 26 measurements from the physical solar cell are collected. Such samples, shown in Table 6.3 represent the experimental data set. The extracted parameters for the SD and DD model are shown in Table 6.2.

Since the SD model has five parameters and DD seven parameters, there are parameters not available for one or other model in Table 6.2. The inexistence of such parameters is represented by the symbol $(-)$. In order to assess the accuracy of the identified model, the relative error R_{error} is defined. R_{error} evaluates the difference between the measured current $I_{t-measured}$ and the calculated by the respective model $I_{t-calculated}$. Therefore, R_{error}, is calculated by:

Table 6.1 Parameter setup for the ABC algorithm

$c_{\max}$	N_p	Limit
10,000	150	$N_p \cdot d$

Table 6.2 Extracted
parameter after applying ABC
for SD and DD

Parameter	Double diode	Single diode
$R_t \Omega$	0.0366	0.0364
$R_{ik} \Omega$	54.3373	53.7185
$I_{pk}(A)$	0.7608	0.7608
$I_{id}(\mu A)$	–	0.3230
$I_{id1}(\mu A)$	0.2238	–
$I_{id2}(\mu A)$	0.2145	–
n_1	1.4551	–
n_2	1.7000	–
n	–	1.4810
RMSE	9.853E−04	9.860E−04

$$R_{error} = \frac{I_{t-measured} - I_{t-calculated}}{I_{t-measured}} \tag{6.12}$$

Table 6.3 presents the extended results of the experiment. Such results include
the experimental data (V_t and I_t), the obtained results ($I_{t-calculted}$) and the respective
relative errors (R_{error}) for both models.

Considering the model parameters of Table 6.2, it is possible to obtain the power
($P = I \times V$) characteristics of the solar cell. Figure 6.3 shows the graphs or current
versus voltage, the power, and the fitness values of the double and single diode
models.

Figure 6.3 presents the results obtained by the presented approach considering
the two diode models. Form Fig. 6.3, it is possible to analyze that the ABC-based
approach obtains accurate models that produce a close approximation to the
experimental data. Besides, the evolution of the optimization process shows that the
presented method allow to find appropriate solar cell models in a reduced number of
generations.

6.5.2   Comparisons with Other Approaches

In order to demonstrate the performance of the presented approach, its results have
been compared to those produced by other similar implementations reported in the
literature, for solar cell modeling. The methods used in the comparison are:
Harmony Search (HS) [13], Particle Swarm Optimization (PSO) [19] and Genetic
Algorithms (GA) [18]. In the comparison, all the algorithms have been executed 35
so that it can be computed their averaged *RMSE* values and their respective standard
deviation (STD). Tables 6.4 and 6.5 present the results obtained from this analysis,
for the DD and SD model, respectively.

Table 6.3 Terminal $(V_t - I_t)$ measurement and relative error values for: double and single diode models

Data	$V_t(V)$ Measured	$I_t(A)$ Measured	$I_{t-calculated}(A)$ ABC double diode model	R_{error} ABC double diode model	$I_{t-calculated}(A)$ ABC single diode model	R_{error} ABC single diode model
1	−0.2057	0.7640	0.7640	−0.0001	0.7641	−0.0001
2	−0.1291	0.7620	0.7626	−0.0008	0.7626	−0.0008
3	−0.0588	0.7605	0.7613	−0.0011	0.7613	−0.0011
4	0.0057	0.7605	0.7601	0.0004	0.7601	0.0004
5	0.0646	0.7600	0.7590	0.0012	0.7590	0.0012
6	0.1185	0.7590	0.7580	0.0012	0.7580	0.0012
7	0.1678	0.7570	0.7571	−0.0001	0.7571	−0.0001
8	0.2132	0.7570	0.7561	0.0011	0.7561	0.0011
9	0.2545	0.7555	0.7550	0.0005	0.7550	0.0005
10	0.2924	0.7540	0.7536	0.0004	0.7536	0.0004
11	0.3269	0.7505	0.7513	−0.0011	0.7513	−0.0011
12	0.3585	0.7465	0.7473	−0.0011	0.7473	−0.0011
13	0.3873	0.7385	0.7401	−0.0021	0.7401	−0.0021
14	0.4137	0.7280	0.7273	0.0008	0.7273	0.0008
15	0.4373	0.7065	0.7069	0.0006	0.7069	−0.0006
16	0.4590	0.6755	0.6752	0.0003	0.6752	0.0003
17	0.4784	0.6320	0.6307	0.0020	0.6307	0.0020
18	0.4960	0.5730	0.5718	0.0019	0.5718	0.0019
19	0.5119	0.4990	0.4995	−0.0011	0.4995	−0.0011
20	0.5265	0.4130	0.4136	−0.0015	0.4136	−0.0015
21	0.5398	0.3165	0.3175	−0.0032	0.3175	−0.0031
22	0.5521	0.2120	0.2121	−0.0008	0.2121	−0.0008
23	0.5633	0.1035	0.1022	0.0118	0.1022	0.0117
24	0.5736	−0.0100	−0.0087	0.1298	−0.0086	0.1315
25	0.5833	−0.1230	−0.1255	−0.0204	−0.1254	−0.0201
26	0.5900	−0.2100	−0.2085	0.0071	−0.2084	0.0073

Table 6.4 presents the comparison analysis for the double diode model. From the results, it is possible to see that the ABC-based algorithm present better performance than other approaches. The STD value can be interpreted as a stability index which reflects the algorithm capacity to produce the same result when it is executed several times. Likewise, Table 6.5 shows the comparison analysis for the single diode model. The results show that the presented algorithm performs better in comparison with the HS, PSO and GA algorithms in terms of the averaged *RMSE* and STD values.

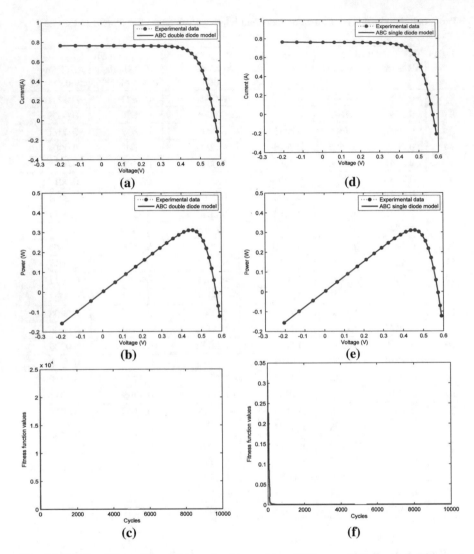

Fig. 6.3 For the DD model: **a** Measured voltage versus ABC computed current, **b** Measured voltage versus ABC-power, **c** *RMSE* evolution. SD model: **d** Measured voltage versus ABC computed current, **e** Measured voltage versus ABC-power, **f** *RMSE* evolution

Recently, statistic tests have been used in several domains [41–46] to validate the performance of new approaches over an experimental data set. Therefore, in order to statistically compare the results from Tables 6.4 and 6.5, a non-parametric significance proof known as the Wilcoxon's rank test [47, 48] for 35 independent samples has been conducted. Such proof allows assessing result differences among two related methods. The analysis is performed considering a 5% significance level over *RMSE* data corresponding to the best solution of each method. Table 6.6

Table 6.4 Comparison results for the DD model

Parameter	ABC	HS	PSO	GA
$R_s(\Omega)$	0.0366	0.0132	0.0000	0.0364
$R_{sh}(\Omega)$	54.3373	57.8142	1.2036	53.7185
$I_{ph}(A)$	0.7608	0.5392	0.7845	0.7608
$I_{sd1}(\mu A)$	0.2238	0.0000	0.0000	0.0000
$I_{sd2}(\mu A)$	0.2145	0.0000	0.0000	0.0000
n_1	1.4551	1.6845	1.7000	1.3355
n_2	1.7000	1.6035	1.7000	1.4810
RMSE	9.853E−04	7.0611	0.2265	9.860E−04
Mean	9.856E−04	6.8348	0.2715	0.0229
STD	3.287E−07	5.1118	0.0309	0.0019

Table 6.5 Comparison results for the SD model

Parameter	ABC	HS	PSO	GA
$R_s(\Omega)$	0.0364	0.0261	0.0000	0.0268
$R_{sh}(\Omega)$	53.7185	32.5141	1.1523	60.0000
$I_{ph}(A)$	0.7608	0.7365	0.8276	0.7635
$I_{sd}(\mu A)$	0.3230	0.0000	0.0000	0.0000
n	1.4810	1.6901	1.7000	1.7000
RMSE	9.860E−04	0.0064	0.2230	0.0047
Mean	0.0001	3.9765	0.2544	0.0015
STD	0.0523	7.2689	0.0289	0.0013

reports the p-values produced by Wilcoxon's test for a pair-wise comparison of the RMSE values between the ABC and the HS, PSO and GA. As a null hypothesis, it is assumed that there is no difference between the values of the two fitness functions. The alternative hypothesis considers an existent difference between the values of both selected approaches. There Wilcoxon's test is performed considering three groups that are formed by ABC versus HS, ABC versus PSO and ABC versus GA, the test is applied over the result obtained for each diode model.

Table 6.6 p-values produced by Wilcoxon's test comparing ABC with HS, PSO and DE

	p-values	
	Double diode model	Single diode model
ABC versus HS	6.5455E−013	1.3246E−011
ABC versus PSO	6.5455E−013	6.5365E−013
ABC versus GA	2.1535E−012	0.0007

All *p*-values reported in the Table 6.6 are less than 0.05 (5% significance level) which is a strong evidence against the null hypothesis, indicating that the *RMSE* values produced by the presented approach are statistically better and they have not occurred by chance.

6.5.3 Robustness

As it is formulated in Eq. 6.6, the objective function is built by using experimental data, extracted from *I–V* measurements of the solar cell. Since experimental data contain noise as a consequence of an imperfect data collection process, the obtained objective function presents high multi-modal and noisy characteristics [28, 29]. Under these circumstances, most of the optimization approaches present a bad performance [30]. In this section, the robustness of the presented approach is tested under different noise conditions. The idea is to evaluate the quality of the generated model when the experimental data are undergone to noise.

The experiment considers the contamination of the experimental data (*I–V*) presented in Table 6.3 by adding Gaussian noise ($\mu = 0, \sigma = 2$). As a consequence of the noisy data, the produced model will present a deterioration in its performance. In order to evaluate the noise sensibility of the produced solar cell models, the averaged deviation D among the noise free (*I–V*) points and the currents (*I–V*) points obtained by the deteriorate model is considered. Therefore, the averaged deviation D is computed as follows:

$$E_i = \sqrt{\left(I_t^i - I_{t-calculated}^i\right)^2 + \left(V_t^i - V_{t-calculated}^i\right)^2}, \quad D = \frac{1}{26}\sum_{i=1}^{26} E_i, \quad (6.13)$$

where E_i represents the deviation for the i (*I–V*) data. I_t^i and V_t^i represent noise free current and voltage, respectively whereas $I_{t-calculated}^i$ and $V_{t-calculated}^i$ corresponds to the current and voltage obtained by the deteriorated model. The resulting D values for the HS, PSO, GA and the presented algorithm are reported in Table 6.7. Figure 6.4 shows the model performance for each algorithm considering noise conditions.

From Table 6.7, it is evident that the presented approach presents the best performance in terms of the averaged deviation D. As a result, the models produced by the ABC-based method presents a minor deterioration than those generated by the other algorithms when they face noise conditions. This fact is attributed to the ABC capacity to solve complex multi-modal objective functions.

Table 6.7 Comparison results of the HS, PSO, GA and the presented algorithm under noise conditions

Point	HS E_i	PSO E_i	GA E_i	ABC E_i
1	0.0327	0.0546	0.0341	0.0042
2	0.0346	0.0218	0.0321	0.0116
3	0.0437	0.0382	0.0540	0.0130
4	0.0459	0.0184	0.0382	0.0127
5	0.0443	0.0505	0.0495	0.0058
6	0.0452	0.0396	0.0393	0.0060
7	0.0398	0.0475	0.0475	0.0070
8	0.0322	0.0450	0.0373	0.0114
6	0.0306	0.0488	0.0386	0.0102
10	0.0292	0.0496	0.0343	0.0046
11	0.0285	0.0488	0.0359	0.0052
12	0.0255	0.0474	0.0373	0.0068
13	0.0271	0.0445	0.0368	0.0112
14	0.0213	0.0417	0.0210	0.0048
15	0.0305	0.0335	0.0329	0.0073
16	0.0181	0.0293	0.0258	0.0073
17	0.0259	0.0122	0.0127	0.0063
18	0.0178	0.0148	0.0162	0.0094
19	0.0180	0.0150	0.0164	0.0046
20	0.0178	0.0191	0.0116	0.0069
21	0.0157	0.0282	0.0134	0.0065
22	0.0144	0.0151	0.0084	0.0082
23	0.0149	0.0149	0.0302	0.0081
24	0.0138	0.0115	0.0104	0.0054
25	0.0125	0.0231	0.0116	0.0061
26	0.0136	0.0156	0.0065	0.0042
D	0.0267	0.0319	0.0282	0.0075

6.6 Conclusions

In this chapter, the use of artificial bee colony (ABC) to accurately estimate the parameter of solar cells has been presented. In the approach, the estimation process is considered as an optimization problem. The presented approach encodes the parameters of the solar cell as a candidate solution. An objective function evaluates the matching quality between a candidate solution and the experimental data. Guided by the values of this objective function, the set of encoded candidate solutions is evolved by using the operators defined by ABC so that the parameters that produce the best possible approximation to the I–V measurements obtained by the true solar cell can be found.

The presented approach has been compared with other similar techniques presented in the literature such as HS, PSO and GA. The efficiency of the algorithm

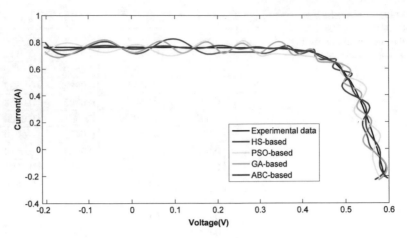

Fig. 6.4 Performance comparison among the optimization algorithms under noisy conditions

has been evaluated in terms of accuracy and robustness. Experimental results provide evidence on the outstanding performance, accuracy and convergence of the presented algorithm in comparison to such methods. The comparisons have been statistically validated considering the Wilcoxon test. Additionally, it has demonstrated that the presented approach is less sensitive to noise conditions than its competitors. As a result, the models produced by the ABC-based method presents a minor deterioration when they face noise conditions. This fact is attributed to the ABC capacity to solve complex multi-modal objective functions.

Although the results offer evidence to demonstrate that the standard ABC method can yield good results on both diode models, the aim of our chapter is not to devise an SC algorithm that could beat all currently available methods, but to show that harmony search algorithms can be effectively considered as an attractive alternative for this purpose.

References

1. Renewables, 2010. Global Status Report. http://www.ren21.net/globalstatusreport/.
2. Ishaque K, Salam Z, Mekhilef S, Shamsudin A. Parameter extraction of solar photovoltaic modules using penalty-based differential evolution. Appl Energy 2012;99:297–308.
3. Orioli A, Gangi AD. A procedure to calculate the five-parameter model of crystalline silicon photovoltaic modules on the basis of the tabular performance data. Appl Energy 2013;102:1160–77.
4. Sandrolini L, Artioli M, Reggiani U. Numerical method for the extraction of photovoltaic module double-diode model parameters through cluster analysis. Appl Energy 2010;87:442–51.
5. Amrouche B, Guessoum A, Belhamel M. A simple behavioural model for solar module electric characteristics based on the first order system step response for MPPT study and comparison. Appl Energy 2012;91:395–404.

6. Bonanno F, Capizzi G, Graditi G, Napoli C, Tina GM. A radial basis function neural network based approach for the electrical characteristics estimation of a photovoltaic module. Appl Energy 2012;97:956–61.

7. L. Han, N. Koide, Y. Chiba, T. Mitate. Modeling of an equivalent circuit for dye-sensitized solar cells. Applied Physics Letters 13 (2004) 2433–2435.

8. M.G. Villalva, J.R. Gazoli, E.R. Filho. Comprehensive approach to modeling and simulation of photovoltaic arrays. IEEE Transactions on Power Electronics 24 (5) (2009) 1198–1208.

9. T. Huld, R. Gottschalg, H.G. Beyer, M. Topic. Mapping the performance of a PV modules, effects of module type and data averaging. Solar Energy 84 (2010) 324–328.

10. W. Xiao, M.G.J Lind, W.G Dunford, A Capel. Real-time identification of optimal operating points in photovoltaic power systems. IEEE Transactions on Industrial Electronics 53 (4) (2006), 1017–1026.

11. M. Chegaar, Z. Ouennough, F. Guechi, H. Langueur. Determination of solar cells parameters under illuminated conditions. Journal of Electron Devices 2 (2003) 17–21.

12. M. Ye, X. Wang, Y. Xu. Parameter extraction of solar cells using particle swarm optimization. Journal of Applied Physics 105 (9) (2009) 094502–094508.

13. Alireza Askarzadeh, Alireza Rezazadeh. Parameter identification for solar cell models using harmony search-based algorithms, Solar Energy 86 (11) (2012) 3241–3249.

14. T. Easwarakhanthan, J. Bottin, I. Bouhouch, C. Boutrit. Nonlinear minimization algorithm for determining the solar cell parameters with microcomputers. Solar Energy (4) 1986 1–12.

15. A. Ortiz-Conde, F.J. Garcia Sanchez, J. Muci. New method to extract the model parameters of solar cells from the explicit analytic solutions of their illuminated I–V characteristics. Solar Energy Materials and Solar Cells 90 (3) (2006) 352–361.

16. D S H Chan, J R Phillips, J C H Phang. A comparative study of extraction methods for solar cell model parameters. Solid-State Electronics 29 (3) (1986) 329–337.

17. M.R. AlRashidi, M.F. AlHajri, K.M. El-Naggar, A.K. Al-Othman. A new estimation approach for determining the I–V characteristics of solar cells. Solar Energy, 85 (7) (2011) 1543–1550.

18. J.A. Jervase, H. Bourdoucen, A. Al-Lawati. Solar cell parameter extraction using genetic algorithms. Measurement Science and Technology 12 (11) (2001) 1922–1925.

19. M. Ye, X. Wang, Y. Xu. Parameter extraction of solar cells using particle swarm optimization. Journal of Applied Physics 105 (9) (2009) 094502–094508.

20. H. Wei, J. Cong, X. Lingyun, S. Deyun. Extracting solar cell model parameters based on chaos particle swarm algorithm. In: International Conference on Electric Information and Control Engineering (ICEICE), (2011) pp. 398–402.

21. K.M. El-Naggar, M.R. AlRashidi, M.F. AlHajri, A.K. Al-Othman. Simulated Annealing algorithm for photovoltaic parameters identification. Solar Energy, 86 (1) (2012) 266–274.

22. Ondřej Hrstka, Anna Kučerová. Improvements of real coded genetic algorithms based on differential operators preventing premature convergence, Advances in Engineering Software, 35, (2004), 237–246.

23. Behrooz OstadmohammadiArani, PooyaMirzabeygi, MasoudShariatPanahi. An improved PSO algorithm with a territorial diversity-preserving scheme and enhanced exploration–exploitation balance, Swarm and Evolutionary Computation, 11, (2013), 1–15.

24. Ling Qing, Wu Gang, Yang Zaiyue, Wang Qiuping. Crowding clustering genetic algorithm for multimodal function optimization, Applied Soft Computing, 8, (2008), 88–95.

25. Minqiang Li, Dan Lin, Jisong Kou. A hybrid niching PSO enhanced with recombination-replacement crowding strategy for multimodal function optimization, Applied Soft Computing, 12, (2012), 975–987.

26. Malihe Niksirat, Mehdi Ghatee, S. Mehdi Hashemi. Multimodal K-shortest viable path problem in Tehran public transportation network and its solution applying ant colony and simulated annealing algorithms, Applied Mathematical Modelling, 36, (2012), 5709–5726.

27. Chia-Ming Wang, Yin-Fu Huang. Self-adaptive harmony search algorithm for optimization, Expert Systems with Applications, 37, (2010), 2826–2837.

28. Jun-hua Li, Ming Li. An analysis on convergence and convergence rate estimate of elitist genetic algorithms in noisy environments, Optik, 124, (2013), 6780–6785.
29. Hui Pan, Ling Wang, Bo Liu. Particle swarm optimization for function optimization in noisy environment, Applied Mathematics and Computation, 181, (2006), 908–919.
30. Hans-Georg Beyer. Evolutionary algorithms in noisy environments: theoretical issues and guidelines for practice, Comput. Methods Appl. Mech. Engrg. 186, (2000), 239–267.
31. D. Karaboga. An idea based on honey bee swarm for numerical optimization, technical report-TR06, Erciyes University, Engineering Faculty, Computer Engineering Department (2005).
32. D. Karaboga, B. Basturk. On the performance of artificial bee colony (ABC) algorithm. Applied Soft Computing, 8 (1) (2008) 687–697.
33. D. Karaboga, B. Akay. A comparative study of artificial bee colony algorithm. Appl Math Comput 214 (2009) 108–132.
34. N. Karaboga. A new design method based on artificial bee colony algorithm for digital IIR filters. J Franklin Inst 346 (2009) 328–348.
35. Q-K. Pan, M. Fatih Tasgetiren, P.N. Suganthan, T.J. Chua. A discrete artificial bee colony algorithm for the lot-streaming flow shop scheduling problem. Information Sciences (2011). doi:10.1016/j.ins.2009.12.025.
36. F. Kang, J. Li, Q. Xu. Structural inverse analysis by hybrid simplex artificial bee colony algorithms. Comput Struct 87 (2009) 861–870.
37. C. Zhang, D. Ouyang, J. Ning. An artificial bee colony approach for clustering. Expert Syst Appl 37 (2010) 4761–4767.
38. D. Karaboga, C. Ozturk. A novel clustering approach: Artificial Bee Colony (ABC) algorithm. Appl Soft Comput 11 (2011) 652–657.
39. S.L. Ho, S. Yang. An artificial bee colony algorithm for inverse problems. Int J Appl Electromagn Mech, 31 (2009) 181–192.
40. D. Karaboga, B. Akay. A comparative study of artificial bee colony algorithm. Appl Math Comput 214 (2009) 108–132.
41. Oliva, D., Cuevas, E., Pajares, G., Zaldivar, D., Osuna, V., A Multilevel thresholding algorithm using electromagnetism optimization, Neurocomputing, (2014), 357–381.
42. Oliva, D., Cuevas, E., Pajares, G., Zaldivar, D., Perez-Cisneros, M., Multilevel thresholding segmentation based on harmony search optimization, Journal of Applied Mathematics, 2013, 575414.
43. Cuevas, E., Zaldivar, D., Pérez-Cisneros, M., Seeking multi-thresholds for image segmentation with Learning Automata, Machine Vision and Applications, 22 (5), (2011), 805–818.
44. Cuevas, E., Ortega-Sánchez, N., Zaldivar, D., Pérez-Cisneros, M., Circle detection by Harmony Search Optimization, Journal of Intelligent and Robotic Systems: Theory and Applications, 66 (3), (2012), 359–376.
45. Cuevas, E., Zaldivar, D., Pérez-Cisneros, M., Ramírez-Ortegón, M., Circle detection using discrete differential evolution Optimization, Pattern Analysis and Applications, 14 (1), (2011), 93–107.
46. Cuevas, E., Echavarría, A., Zaldívar, D., Pérez-Cisneros, M., A novel evolutionary algorithm inspired by the states of matter for template matching, Expert Systems with Applications, 40 (16), (2013), 6359–6373.
47. Wilcoxon, F.: Individual comparisons by ranking methods. Biometrics 1, 80–83 (1945).
48. Garcia, S., Molina, D., Lozano, M.,Herrera, F.:A study on the use of non-parametric tests for analyzing the evolutionary algorithms' behaviour: a case study on the CEC'2005 Special session on real parameter optimization. J. Heurist. (2008). doi:10.1007/s10732-008-9080-4.

Chapter 7
Parameter Identification of Induction Motors

Abstract The efficient use of electrical energy is a topic that has attracted attention for its environmental consequences. On the other hand, induction motors represent the main component in most of the industries. They consume the highest energy percentages in industrial facilities. This energy consumption depends on the operation conditions of the induction motor imposed by its internal parameters. Since the internal parameters of an induction motor are not directly measurable, an identification process must be conducted to obtain them. In the identification process, the parameter estimation is transformed into a multidimensional optimization problem where the internal parameters of the induction motor are considered as decision variables. Under this approach, the complexity of the optimization problem tends to produce multimodal error surfaces for which their cost functions are significantly difficult to minimize. Several algorithms based on evolutionary computation principles have been successfully applied to identify the optimal parameters of induction motors. However, most of them maintain an important limitation, they frequently obtain sub-optimal solutions as a result of an improper equilibrium between exploitation and exploration in their search strategies. This chapter presents an algorithm for the optimal parameter identification of induction motors. To determine the parameters, the presented method uses a recent evolutionary method called the Gravitational Search Algorithm (GSA). Different to the most of existent evolutionary algorithms, GSA presents a better performance in multimodal problems, avoiding critical flaws such as the premature convergence to sub-optimal solutions. Numerical simulations have been conducted on several models to show the effectiveness of the presented scheme.

7.1 Introduction

The environmental consequences that overconsumption of electrical energy entails has recently attracted the attention in different fields of the engineering. Therefore, the improvement of machinery and elements that have high electrical energy consumptions has become an important task nowadays [1].

Induction motors present several benefits such as their ruggedness, low price, cheap maintenance and easy operation [2]. However, more than a half of electric energy consumed by industrial facilities is due to use of induction motors. With the massive use of induction motors, the electrical energy consumption has increased exponentially through the years. This fact has generated the need to improve their efficiency which mainly depends on their internal parameters. The parameter estimation of induction motors represents a complex task due to its non-linearity. As a consequence, different alternatives have been presented in the literature. Some examples include the proposed by Waters and Willoughby [3], where the parameter are estimated from the knowledge of certain variables such as stator resistance and the leakage reactance, the proposed by Ansuj [4], where the identification is based on a sensitivity analysis and the proposed by De Kock [5], where the estimation is conducted through an output error technique.

As an alternative to such techniques, the problem of parameter estimation in induction motors has also been handled through evolutionary methods. In general, they have demonstrated, under several circumstances, to deliver better results than those based on deterministic approaches in terms of accuracy and robustness [6]. Some examples of these approaches used in the identification of parameters in induction motors involve methods such as Genetic Algorithms (GA) [7], Particle Swarm Optimization (PSO) [8, 9], Artificial immune system (AIS) [10], Bacterial Foraging Algorithm (BFA) [11], Shuffled Frog-Leaping algorithm [12], hybrid of genetic algorithm and particle swarm optimization [8], multiple-global-best guided artificial bee colony [9], just to mention a few. Although these algorithms present interesting results, they have an important limitation: They frequently obtain sub-optimal solutions as a consequence of the limited balance between exploration and exploitation in their search strategies.

On the other hand, the Gravitational Search Algorithm (GSA) [13] is a recent evolutionary computation algorithm which is inspired on the physical phenomenon of the gravity. In GSA, its evolutionary operators are built considering the gravitation principles. Different to most of existent evolutionary algorithms, GSA presents a better performance in multimodal problems, avoiding critical flaws such as the premature convergence to sub-optimal solutions [14, 15]. Such characteristics have motivated its use to solve an extensive variety of engineering applications such as energy [16], image processing [6] and machine learning [17].

This chapter presents an algorithm for the optimal parameter identification of induction motors. To determine the parameters, the presented method uses a recent evolutionary method called the Gravitational Search Algorithm (GSA). A comparison with state-of-the-art methods such as Artificial Bee Colony (ABC) [18], Differential Evolution (DE) [19] and Particle Swarm Optimization (PSO) [20] on different induction models has been incorporated to demonstrate the performance of the presented approach. Conclusions of the experimental comparison are validated through statistical tests that properly support the discussion.

The sections of this chapter are organized as follows: Sect. 7.2 describes the GSA method. In Sect. 7.3 the identification problem is exposed. In Sect. 7.4, the experimental results are presented. Finally, in Sect. 7.5 the conclusions are stated.

7.2 Gravitational Search Algorithm

The Gravitational Search Algorithm (GSA) was firstly introduced by Rashedi [13] in 2009 inspired on the laws of gravity. Different to most of existent evolutionary algorithms, GSA presents a better performance in multimodal problems, avoiding critical flaws such as the premature convergence to sub-optimal solutions [14, 15]. In GSA, candidate solutions emulate masses which attract to each other through operators that mimic the gravitational force. Under GSA, the mass (quality) of each candidate solution is assigned according to correspondent fitness value. GSA has been designed to find the global solution of a nonlinear optimization problem with box constraints in the form:

$$\begin{array}{ll} \text{maximize} & f(\mathbf{x}),\ \mathbf{x} = (x_1, \ldots, x_d) \in \mathbb{R}^d \\ \text{subject to} & \mathbf{x} \in \mathbf{X} \end{array} \tag{7.1}$$

where $f : \mathbb{R}^d \to \mathbb{R}$ is a nonlinear function whereas $\mathbf{X} = \{\mathbf{x} \in \mathbb{R}^d \,|\, l_h \leq x^h \leq u_h, h = 1, \ldots, d\}$ is a bounded feasible space, constrained by the lower (l_h) and upper (u_h) limits. To solve the problem formulated in Eq. 7.1, GSA utilizes a population of N candidate solutions. Each mass (or candidate solution) represents a d-dimensional vector $\mathbf{x}_i(t) = (x_i^1, \ldots, x_i^d)$ ($i \in 1 \ldots, N$), where each dimension correspond to a decision variable of the optimization problem at hand.

In GSA, at a time t the force acting from a mass i to a mass j of the h variable ($h \in 1 \ldots, d$) is defined as follows:

$$F_{ij}^h(t) = G(t) \frac{Mp_i(t) \times Ma_j(t)}{R_{ij}(t) + \varepsilon} (x_j^h(t) - x_i^h(t)) \tag{7.2}$$

where Ma_j is the active gravitational mass related to solution j and Mp_i symbolizes the passive gravitational mass of solution i, $G(t)$ is the gravitational constant at time t, ε is a small constant and R_{ij} is the Euclidean distance between the i-th and j-th individuals. In GSA, $G(t)$ is a function which is modified during the evolution process. The idea with this modification is to adjust the balance between exploration and exploitation through the alteration of the attraction forces among solutions.

The total force acting over a candidate solution i is defined by the following model:

$$F_i^h(t) = \sum_{j=1, j \neq i}^{N} F_{ij}^h(t) \tag{7.3}$$

Then, the acceleration of the candidate solution i at time t is computed as follows:

$$a_i^h(t) = \frac{F_i^h(t)}{Mn_i(t)} \tag{7.4}$$

where Mn_i represents the inertial mass of the candidate solution i. Under such conditions, the new position of each candidate solution i is calculated as follows:

$$\begin{aligned}
x_i^h(t+1) &= x_i^h(t) + v_i^h(t+1) \\
v_i^h(t+1) &= \text{rand}() \cdot v_i^h(t) + a_i^h(t)
\end{aligned} \tag{7.5}$$

At each iteration, the gravitational and inertia masses of each particle are evaluated in terms of its fitness function. Therefore, the gravitational and inertia masses are updated by the following equations:

$$Ma_i = Mp_i = M_{ii} = M_i \tag{7.6}$$

$$m_i(t) = \frac{f(\mathbf{x}_i(t)) - worst(t)}{best(t) - worst(t)} \tag{7.7}$$

$$M_i(t) = \frac{m_i(t)}{\sum_{j=1}^{N} m_j(t)}, \tag{7.8}$$

where $f(\cdot)$ represents the objective function whose final result exhibits the fitness value. On the other hand, best(t) and worst(t) symbolizes the best and worst fitness values found at time t in the complete population. Figure 7.1 illustrates the pseudo code of the GSA method.

1.   Random Initialization of the population
2.   Find the best and worst solutions in the initial population
3.   **while** (*stop criteria*)
4.   **for** *i=1:N* (for all elements)
5.   update $G(t)$, *best(t)*, *worst(t)* and $M_i(t)$ for $i = 1, 2.., N$
6.   calculate the mass of individual $M_i(t)$
7.   calculate the gravitational constant $G(t)$,
8.   calculate acceleration $a_i^h(t)$
9.   update the velocity and positions v_i^h, x_i^h
10.   **end for**
11.   Find the best individual
12. **end while**
13. Display the best individual as the solution

Fig. 7.1 Gravitational search algorithm (GSA) pseudo code

7.3   Identification Problem Formulation

The parameters of an induction motor are not directly measurable. Due to this, they are commonly estimated by identification methods. Under such approaches, the behavior of an induction motor is modeled by equivalent nonlinear circuits. Depending on the accuracy, there exist two different circuit models [10]: The approximate circuit model and the exact circuit model. In general, they allow the adequate relation of the motor parameters for their estimation.

In the identification process, the parameter estimation is transformed into a multidimensional optimization problem where the internal parameters of the induction motor are considered as decision variables. Therefore, the objective is to minimize the error between the estimated and the manufacturer data adjusting the parameters of the equivalent circuit. Under this approach, the complexity of the produced formulations tends to produce multimodal error surfaces for which their cost functions are significantly difficult to minimize.

7.3.1   Approximate Circuit Model

The approximate circuit model does not consider the magnetizing reactance and rotor reactance in its structure; hence, its accuracy is less than the exact circuit model. The approximate circuit model uses the manufacturer data starting torque (T_{lr}), maximum torque (T_{max}) and full load torque (T_{fl}) to determinate the stator resistance (R_1), rotor resistance (R_2), stator leakage reactance (X_1) and motor slip (s). Figure 7.2 illustrates the approximate circuit model. Under the approximate circuit model, the identification task can be formulated as the following optimization problem:

$$
\begin{aligned}
\text{maximize}\quad & f(\mathbf{x}), \quad \mathbf{x} = (x_1, \ldots, x_d) \in \mathbb{R}^d \\
\text{subject to}\quad & 0 \le R_1 \le 1 \\
& 0 \le R_2 \le 1 \\
& 0 \le X_1 \le 10 \\
& 0 \le s \le 1
\end{aligned}
\tag{7.9}
$$

Fig. 7.2 Approximate circuit model

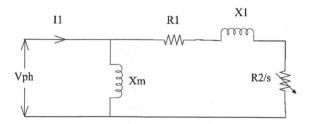

where:

$$J_A(\mathbf{x}) = (f_1(\mathbf{x}))^2 + (f_2(\mathbf{x}))^2 + (f_3(\mathbf{x}))^2$$

$$f_1(\mathbf{x}) = \frac{\dfrac{K_t R_2}{s\left[(R_1 + R_2/s)^2 + X_1^2\right]} - T_{fl}}{T_{fl}}$$

$$f_2(\mathbf{x}) = \frac{\dfrac{K_t R_2}{(R_1 + R_2)^2 + X_1^2} - T_{lr}}{T_{lr}}$$

$$f_3(\mathbf{x}) = \frac{\dfrac{K_t}{2\left[R_1 + \sqrt{R_1^2 + X_1^2}\right]} - T_{\max}}{T_{\max}}$$

(7.10)

$$K_t = \frac{3V_{ph}^2}{\omega_s}$$

7.3.2   Exact Circuit Model

Different to the approximate circuit model, in the exact circuit model, the effects of the magnetizing reactance and rotor reactance are considered in the computation. In this model, it is calculated the stator resistance (R_1), rotor resistance (R_2), stator leakage inductance (X_1), rotor leakage reactance (X_2), magnetizing leakage reactance (X_m) and motor slip (s) to determinate the maximum torque $(T_{\max})$, full load torque (T_{fl}), starting torque (T_{str}) and full load power factor (pf). Figure 7.3 shows the exact circuit model. Under the exact circuit model, the identification task can be formulated as the following optimization problem:

$$
\begin{aligned}
\text{minimize} \quad & J_E(\mathbf{x}), \quad \mathbf{x} = (R_1, R_2, X_1, X_2, X_m, s) \in \mathbb{R}^6 \\
\text{subject to} \quad & 0 \le R_1 \le 1 \\
& 0 \le R_2 \le 1 \\
& 0 \le X_1 \le 1 \\
& 0 \le X_2 \le 1 \\
& 0 \le X_m \le 10 \\
& 0 \le s \le 1
\end{aligned}
$$

(7.11)

Fig. 7.3 Exact circuit model

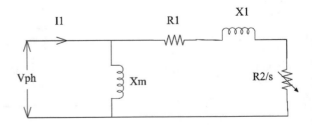

where

$$J_E(\mathbf{x}) = (f_1(\mathbf{x}))^2 + (f_2(\mathbf{x}))^2 + (f_3(\mathbf{x}))^2 + (f_4(\mathbf{x}))^2$$

$$f_1(\mathbf{x}) = \frac{\frac{K_t R_2}{s\left[(R_{th}+R_2/s)^2+X^2\right]} - T_{fl}}{T_{fl}}$$

$$f_2(\mathbf{x}) = \frac{\frac{K_t R_2}{(R_{th}+R_2)^2+X^2} - T_{str}}{T_{str}}$$

$$f_3(\mathbf{x}) = \frac{\frac{K_t}{2\left[R_{th}+\sqrt{R_{th}^2+X^2}\right]} - T_{\max}(mf)}{T_{\max}(mf)} \qquad (7.12)$$

$$f_4(\mathbf{x}) = \frac{\cos\left(\tan^{-1}\left(\frac{X}{R_{th}+R_2/s}\right)\right) - pf}{pf}$$

$$V_{th} = \frac{V_{ph}X_m}{X_1+X_m}, R_{th} = \frac{R_1 X_m}{X_1+X_m}, X_{th} = \frac{X_1 X_m}{X_1+X_m},$$

$$K_t = \frac{3V_{th}^2}{\omega_s}, X = X_2 + X_{th}$$

In the minimization of Eq. 7.11, it is also necessary to meet an additional condition, the values of the calculated parameters must fulfill the following restriction:

$$\frac{p_{fl} - (I_1^2 R_1 + I_2^2 R_2 + P_{rot})}{p_{fl}} = \eta_{fl} \qquad (7.13)$$

where P_{rot} represents the rotational power losses, η_{fl} symbolizes the efficiency and p_{fl} is the rated power.

7.4   Experimental Results

In this chapter, the Gravitational search algorithm (GSA) is used to determine the optimal parameters of two induction motors considering the approximate circuit model (J_A) and exact circuit model (J_E). Table 7.1 presents the technical characteristics of both motors used in the experiments. The presented method is also evaluated in comparison with other similar approaches based on evolutionary algorithms. In the experiments, we have applied the GSA-estimator to the parameter identification of both induction motors whereas its results are compared to those produced by Differential evolution (DE) [20], artificial bee colony (ABC) [18] and Particle Swarm Optimization (PSO) [19]. The parameter settings of all compared algorithms are obtained from their own referenced papers. The parameter setting for each algorithm in the comparison is described as follows:

Table 7.1 Manufacturer data
of the motors used in the
experiments

	Motor 1	Motor 2
Power (HP)	5	40
Voltage (V)	400	400
Current (A)	8	45
Frequency (Hz)	50	50
No. Poles	4	4
Full load slip (s)	0.07	0.09
Starting torque (T_{str})	15	260
Max. Torque (T_{max})	42	370
Stator current	22	180
Full load torque (T_{fl})	25	190

1. PSO, parameters $c_1 = 2$, $c_2 = 2$ and weights factors were set $w_{max} = 0.9$, and $w_{min} = 0.4$ [19].
2. ABC, the parameters implemented were provided by [18], limit = 100.
3. DE, in accordance with [20] the parameters were set $p_c = 0.5$ and $f = 0.5$
4. GSA, the parameter were set according to [13].

The experimental results are divided into three sub-sections. In the first Sect. (6.4.1), the performance of the presented algorithm is evaluated with regard to its own tuning parameters (sensibility analysis). In the second Sect. (7.4.2), an overall performance of the presented method in comparison with similar approaches is provided. Finally, in the third Sect. (7.4.3), the results are statistically analyzed and validated by using the Wilcoxon test.

7.4.1 Performance Evaluation with Regard to Its Own Tuning Parameters

In GSA, the parameters G_0 and α affect mainly its expected performance [R]. In this sub-section, it is analyzed the behavior of the GSA over the motor parameter estimation problem considering different setting parameters.

During the test, each parameter G_0 and α is set to a default value such as $G_0 = 100$ and $\alpha = 20$. In the analysis, when one of the two parameters is evaluated, the other parameter remain fixed to the default value. To minimize the stochastic effect of the algorithm, each benchmark function is executed independently 30 times. As a termination criteria, it is considered the maximum number of iterations which has been set to 3000. In all simulations, the population size N has been configured to 25 individuals.

Table 7.2 Experimental results obtained by the presented algorithm using different values of G_0

	$G_0 = 80$ $\alpha = 20$	$G_0 = 90$ $\alpha = 20$	$G_0 = 100$ $\alpha = 20$	$G_0 = 110$ $\alpha = 20$	$G_0 = 120$ $\alpha = 20$
Min	0.0044	0.0036	0.0032	0.0036	0.0033
Max	0.0119	0.0103	0.0032	0.0082	0.0088
Std	0.0016	0.0013	0.0000	0.0012	0.0014
Mean	0.0052	0.0040	0.0032	0.0042	0.0039

In the first stage, the behavior of the presented algorithm is analyzed considering different values for G_0. In the analysis, the values of G_0 are varied from 80 to 120 whereas the values of α remains fixed to 10 and 30, respectively. In the simulation, the presented method is executed independently 30 times for each value of G_0. The results obtained for the parameter combination of G_0 and α are shown in Table 6.2. Such values represent the minimum, maximum, standard deviation and mean values of J_E (exact circuit model), considering the characteristics of motor 1. The best results are marked in boldface. From Table 7.2, we can conclude that the presented GSA algorithm with $G_0 = 100$ maintains the best performance.

Then, in the second stage, the performance of the presented algorithm is evaluated considering different values for α. In the experiment, the values of α are varied from 10 to 30 whereas the value of G_0 remains fixed to 100. The statistical results obtained by the GSA algorithm using different values of α are presented in Table 7.3. Such values represent the minimum, maximum, standard deviation and mean values of J_E (exact circuit model), considering the characteristics of motor 2. The best results are marked in boldface. From Table 7.3, it is evident that the presented algorithm with $\alpha = 20$ outperforms the other parameter configurations.

In general, once observed Tables 7.2 and 7.3, the experimental results suggest that a proper combination of different parameter values can improve the performance of the presented method and the quality of solutions. With the experiment can be concluded that the best parameter set is composed by the following values: $G_0 = 100$ and $\alpha = 20$. Once they have been determined experimentally, they are kept for all the test functions through the next experiments.

Table 7.3 Experimental results obtained by the presented algorithm using different values of α

	$G_0 = 100$ $\alpha = 10$	$G_0 = 100$ $\alpha = 15$	$G_0 = 100$ $\alpha = 20$	$G_0 = 100$ $\alpha = 25$	$G_0 = 100$ $\alpha = 30$
Min	0.0093	0.0093	0.0071	0.0093	0.0092
Max	0.0730	0.0433	0.0209	0.0435	0.0493
Std	0.0147	0.0085	0.0043	0.0094	0.0109
Mean	0.0235	0.0164	0.0094	0.0191	0.0215

7.4.2   *Induction Motor Parameter Identification*

In this experiment, the performance of the presented GSA method is compared with DE, ABC and PSO, considering the parameter estimation of both circuit models. In the test, all algorithms are operated with a population of 25 individuals (N = 25). The maximum iteration number for all methods has been set to 3000. This stop criterion has been selected to maintain compatibility to similar works reported in the literature [19–20]. All the experimental results presented in this section consider the analysis of 35 independent executions of each algorithm. Thus, the values of J_A (approximate model), deviation standard and mean obtained by each algorithm for the motor 1 are reported in Table 7.4 whereas the results produced by the motor 2 are shown in Table 7.5. On the other hand, the values of J_E (exact model) for motor 1 and motor 2 are exhibited in Tables 7.6 and 7.7, respectively. The best results in all Tables are marked in boldface.

According to the results from Tables 7.4, 7.5, 7.6 and 7.7, the presented approach provides better performance than DE, ABC and PSO for all tests. These differences are directly related to a better trade-off between exploration and exploitation of the GSA method.

Table 7.4 Results of J_A considering motor 1

	GSA	DE	ABC	PSO
Min	3.4768e−22	1.9687e−15	2.5701e−05	1.07474e−04
Max	1.6715e−20	0.0043	0.0126	0.0253
Mean	5.4439e−21	1.5408e−04	0.0030	0.0075
Std	4.1473e−21	7.3369e−04	0.0024	0.0075

Table 7.5 Results of J_A considering motor 2

	GSA	DE	ABC	PSO
Min	3.7189e−20	1.1369e−13	3.6127e−04	0.0016
Max	1.4020e−18	0.0067	0.0251	0.0829
Mean	5.3373e−19	4.5700e−04	0.0078	0.0161
Std	3.8914e−19	0.0013	0.0055	0.0165

Table 7.6 Results of J_E considering motor 1

	GSA	DE	ABC	PSO
Min	0.0032	0.0172	0.0172	0.0174
Max	0.0032	0.0288	0.0477	0.0629
Mean	0.0032	0.0192	0.0231	0.0330
Std	0.0000	0.0035	0.0103	0.0629

Table 7.7 Results of J_E considering motor 2

	GSA	DE	ABC	PSO
Min	0.0071	0.0091	0.0180	0.0072
Max	0.0209	0.0305	0.2720	0.6721
Mean	0.0094	0.0190	0.0791	0.0369
Std	0.0043	0.0057	0.0572	0.1108

Once estimated the motor parameters of all algorithms, their estimations are compared with the ideal starting torque (T_{str}), Max. Torque (T_{max}) and full load torque (T_{fl}) values provided by the manufacturer in Table 6.1. The main objective with this comparison is to evaluate the accuracy of each approach with regard to the actual motor parameters. Tables 6.8 and 6.9 present the experimental results of J_A for motors 1 and 2, respectively. On the other hand, Tables 7.10 and 7.11 exhibits the comparative results of J_E for motors 1 and 2, respectively. The best results in all Tables are marked in boldface (Tables 7.8 and 7.9).

Since the convergence rate of evolutionary algorithms is an important characteristic to assess their performance for solving the optimization problems, the convergence of all algorithms facing functions J_A and J_E is compared in Fig. 7.4a, b. The remarkable convergence rate of the presented algorithm can be observed from

Table 7.8 Comparison of GSA, DE, ABC and PSO with manufacturer data, J_A, motor 1

	True value	GSA	Error %	DE	Error %	ABC	Error %	PSO	Error %
Tst	15	15.00	0	14.9803	−0.131	14.3800	−4.133	15.4496	2.9973
Tmax	42	42.00	0	42.0568	0.135	40.5726	−3.398	39.6603	−5.570
Tfl	25	25.00	0	24.9608	−0.156	25.0480	0.192	25.7955	3.182

Table 7.9 Comparison of GSA, DE, ABC and PSO with manufacturer data, J_A, motor 2

	True value	GSA	Error %	DE	Error %	ABC	Error %	PSO	Error %
Tst	260	260.00	0	258.4709	−0.588	260.6362	0.2446	288.9052	11.117
Tmax	370	370.00	0	372.7692	0.7484	375.0662	1.3692	343.5384	−7.151
Tfl	190	190.00	0	189.0508	−0.499	204.1499	7.447	196.1172	3.2195

Table 7.10 Comparison of GSA, DE, ABC and PSO with manufacturer data, J_E, motor 1

	True value	GSA	Error %	DE	Error %	ABC	Error %	PSO	Error %
Tst	15	14.9470	−0.353	15.4089	2.726	16.4193	9.462	15.6462	4.308
Tmax	42	42.00	0	42.00	0	42.00	0	42.00	0
Tfl	25	25.0660	0.264	26.0829	4.3316	25.3395	1.358	26.6197	6.4788

Table 7.11 Comparison of GSA, DE, ABC and PSO with manufacturer data, J_E, motor 2

	True value	GSA	Error %	DE	Error %	ABC	Error %	PSO	Error %
Tst	260	258.1583	−0.708	262.0565	0.7909	246.2137	−5.302	281.8977	8.4221
Tmax	370	370.00	0	370.00	0	370.00	0	370.00	0
Tfl	190	189.8841	−0.061	192.2916	1.2061	207.9139	9.428	166.6764	−12.27

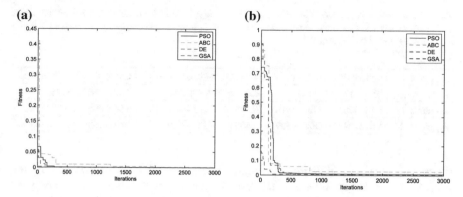

Fig. 7.4 Convergence evolution through iterations: **a** Model 1(J_A), **b** Model 2 (J_E)

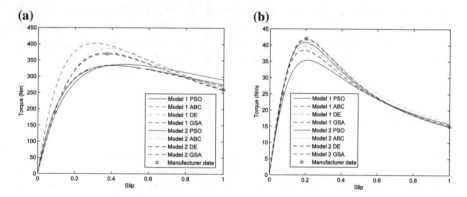

Fig. 7.5 Curve slip vs. torque using PSO, ABC, DE and GSA considering Model 1(J_A), Model 2 (J_E), **a** Motor 1, **b** Motor 2

both figures. According to these figures, it tends to find the global optimum faster than other algorithms.

Finally, Fig. 7.5 shows graphically the relation of the slip versus torque for both models (J_A and J_E) and for both motors (1 and 2) (Tables 7.8 and 7.9).

7.4.3 Statistical Analysis

To statistically analyze the results, a non-parametric test known as the Wilcoxon analysis [21] has been conducted. It permits to evaluate the differences between two related methods. The test is performed in a 5% of significance level over the mean

Table 7.12 p-values produced by Wilcoxon test comparing GSA versus DE, GSA versus ABC and GSA versus PSO over the mean fitness values of J_A and $J_B J_B$ considering the motors 1 and 2 from Tables 7.4, 7.5, 7.6 and 7.7

GSA vs.	DE	ABC	PSO
J_A, motor 1	6.545500588914223e−13	6.545500588914223e−13	6.545500588914223e−13
J_A, motor 2	0.009117078811112	0.036545600995029	0.004643055264741
J_E, motor 1	6.545500588914223e−13	6.545500588914223e−13	6.545500588914223e−13
J_E, motor 2	1.612798082388261e−09	9.465531545379272e−13	3.483016312301559e08−

fitness values of J_A and J_E considering the motors 1 and 2. Table 7.12 reports the p-values generated by Wilcoxon analysis for the pair-wise comparison among the algorithms. Under such conditions, three groups are produced: GSA vs. DE, GSA vs. ABC and GSA vs. PSO. In the Wilcoxon test, it is assumed as a null hypothesis that there is no significant difference between the two algorithms. On the other hand, it is considered as alternative hypothesis that there is a significant difference between both approaches. An inspection of Table 7.12 demonstrates that all p-values in the Table 7.12 are less than 0.05 (5% significance level). This fact gives a strong evidence against the null hypothesis, indicating that the presented method statistically presents better results than the other algorithms.

7.5 Conclusions

In this chapter, an algorithm for the optimal parameter identification of induction motors has been presented. In the presented method, the parameter estimation process is transformed into a multidimensional optimization problem where the internal parameters of the induction motor are considered as decision variables. Under this approach, the complexity of the optimization problem tends to produce multimodal error surfaces for which their cost functions are significantly difficult to minimize. To determine the parameters, the presented method uses a relatively recent evolutionary method called the Gravitational Search Algorithm (GSA). Different to most of existent evolutionary algorithms, GSA presents a better performance in multimodal problems, avoiding critical flaws such as the premature convergence to sub-optimal solutions.

To illustrate the proficiency and robustness of the presented approach, the GSA-estimator has been experimentally evaluated considering two different motor models. To assess the performance of the presented algorithm, it has been compared to other similar evolutionary approaches such as differential evolution (DE), artificial bee colony (ABC) and particle swarm optimization (PSO). The experiments, statistically validated, have demonstrated that presented method outperforms the other techniques for most of the experiments in terms of solution quality.

References

1. H. Çaliş, A. Çakir, and E. Dandil, "Artificial immunity-based induction motor bearing fault diagnosis," *Turkish J. Electr. Eng. Comput. Sci.*, vol. 21, no. 1, pp. 1–25, 2013.
2. V. Prakash, S. Baskar, S. Sivakumar, and K. S. Krishna, "A novel efficiency improvement measure in three-phase induction motors, its conservation potential and economic analysis," *Energy Sustain. Dev.*, vol. 12, no. 2, pp. 78–87, 2008.
3. S. S. Waters and R. D. Willoughby, "Modeling Induction Motors for System Studies," *IEEE Trans. Ind. Appl.*, vol. IA-19, no. 5, pp. 875–878, 1983.
4. S. Ansuj, F. Shokooh, and R. Schinzinger, "Parameter estimation for induction machines based on sensitivity\nanalysis," *IEEE Trans. Ind. Appl.*, vol. 25, no. 6, pp. 1035–1040, 1989.
5. J. De Kock, F. Van der Merwe, and H. Vermeulen, "Induction motor parameter estimation through an output error technique," *Energy Conversion, IEEE Trans.*, vol. 9, no. 1, pp. 69–76, 1994.
6. V. Kumar, J. K. Chhabra, and D. Kumar, "Automatic cluster evolution using gravitational search algorithm and its application on image segmentation," *Eng. Appl. Artif. Intell.*, vol. 29, pp. 93–103, 2014.
7. R. R. Bishop and G. G. Richards, "Identifying induction machine parameters using a genetic optimization algorithm", Southeastcon '90. Proceedings., IEEE pp. 476–479, 1990.
8. H. R. Mohammadi and A. Akhavan, "Parameter Estimation of Three-Phase Induction Motor Using Hybrid of Genetic Algorithm and Particle Swarm Optimization," vol. 2014, 2014.
9. A. G. Abro and J. Mohamad-Saleh, "Multiple-global-best guided artificial bee colony algorithm for induction motor parameter estimation," *Turkish J. Electr. Eng. Comput. Sci.*, vol. 22, pp. 620–636, 2014.
10. V. P. Sakthivel, R. Bhuvaneswari, and S. Subramanian, "Artificial immune system for parameter estimation of induction motor," *Expert Syst. Appl.*, vol. 37, no. 8, pp. 6109–6115, 2010.
11. V. P. Sakthivel, R. Bhuvaneswari, and S. Subramanian, "An accurate and economical approach for induction motor field efficiency estimation using bacterial foraging algorithm," *Meas. J. Int. Meas. Confed.*, vol. 44, no. 4, pp. 674–684, 2011.
12. I. Perez, M. Gomez-Gonzalez, and F. Jurado, "Estimation of induction motor parameters using shuffled frog-leaping algorithm," *Electr. Eng.*, vol. 95, no. 3, pp. 267–275, 2013.
13. E. Rashedi, H. Nezamabadi-pour, and S. Saryazdi, "GSA: A Gravitational Search Algorithm," *Inf. Sci. (Ny).*, vol. 179, no. 13, pp. 2232–2248, 2009.
14. F. Farivar and M. A. Shoorehdeli, "Stability analysis of particle dynamics in gravitational search optimization algorithm," *Inf. Sci. (Ny).*, vol. 337–338, pp. 25–43, 2016.
15. S. Yazdani, H. Nezamabadi-Pour, and S. Kamyab, "A gravitational search algorithm for multimodal optimization," *Swarm Evol. Comput.*, vol. 14, pp. 1–14, 2014.
16. A. Yazdani, T. Jayabarathi, V. Ramesh, and T. Raghunathan, "Combined heat and power economic dispatch problem using firefly algorithm," *Front. Energy*, vol. 7, no. 2, pp. 133–139, 2013.
17. W. Zhang, P. Niu, G. Li, and P. Li, "Forecasting of turbine heat rate with online least squares support vector machine based on gravitational search algorithm," *Knowledge-Based Syst.*, vol. 39, pp. 34–44, 2013.
18. M. Jamadi and F. Merrikh-bayat, "New Method for Accurate Parameter Estimation of Induction Motors Based on Artificial Bee Colony Algorithm."
19. V. P. Sakthivel and S. Subramanian, "On-site efficiency evaluation of three-phase induction motor based on particle swarm optimization," *Energy*, vol. 36, no. 3, pp. 1713–1720, 2011.

20. R. K. Ursem and P. Vadstrup, "Parameter identification of induction motors using differential evolution," *Evol. Comput. 2003. CEC'03. 2003 Congr.*, vol. 2, pp. 790–796 Vol. 2, 2003.
21. F. Wilcoxon, "Breakthroughs in Statistics: Methodology and Distribution," S. Kotz and N. L. Johnson, Eds. New York, NY: Springer New York, 1992, pp. 196–2.

Chapter 8
White Blood Cells Detection in Images

Abstract As a research area, there are several problems in medical imaging that continue unresolved; one of those is the automatic detection of white blood cells (WBC) in smear images. The study of this kind of images has engaged researchers from fields of medicine and computer vision alike. Several studies have been done to try to approximate this cells with circular or ellipsoid forms; once detected, those cells can be further processed by computer vision systems. In this chapter, detection of WBC in smear digitalized images is achieved by using evolutionary algorithms, with an objective function that considers that since WBC can be approximated by an ellipsoid form, an ellipse detector algorithm may be successfully applied in order to recognize them. In that sense, the optimization problem also consider that a candidate solution is a probable ellipse that could adjust a WBC in the image.

8.1   Introduction

The utilization of medical images (MI) by physicians has been the cornerstone in medicine for centuries; however, the traditional analysis of them by an expert is prone to error, because of several factors, which include tiredness, complicated images, distractions, etc., and also, is time consuming [1]. For those reasons, since some decades ago has become common the use of computers to retrieve, store, and process this kind of images. The main idea behind the utilization of those tools is to achieve better, faster, and reliable, both diagnostics along with treatment of diseases. Those are commonly called computer assisted diagnosis (CAD) systems [1].

As the quality of images increases, at the same time are required better algorithms, as well as bigger storage spaces to analyze and process those amounts of digital information [2], being the majority of them based on parallelization of the analysis routines, or based on unconventional computing [3]. There are several ways to obtain MI: X-ray radiography, magnetic resonance, ultrasound, and optically obtained images. Belonging to the last category is the optical retrieving of smear samples, which are later digitalized and used by some fully or partially automated image processing system. Depending on the tissue being analyzed, the

© Springer International Publishing AG 2017
E. Cuevas et al., *Evolutionary Computation Techniques:
A Comparative Perspective*, Studies in Computational Intelligence 686,
DOI 10.1007/978-3-319-51109-2_8

images could correspond to papanicolau test [4], urine [5], blood [6], etc. Regarding to blood, there are several diseases that could be discovered through its analysis; white blood cells, or leukocytes, are a key marker for the diagnosis of some kind of cancer in that tissue.

Many computer vision techniques have successfully contributed to generate new methods for cell analysis, leading to more effective computer vision systems, and therefore to better diagnoses. Nevertheless, there exists several variables that complicates the data extraction process, such as high variability on cell shape, size, edge and localization. As the samples are prepared by following a manual process, the obtained images could have poor contrast, high variability on cells shape, edge, size, and location, provoking problems to the automatic image processing system. Several researches have been done in the area of WBC detection, and more specifically for segmentation. For example, the use of boundary support vectors was proposed in [7], which are obtained directly from 1D histogram, and also used to make the segmentation of the three classes in images. After that operation, authors propose the use of morphological operations in order to get more defined segmented WBC in the detection process. An automated approach, that uses simple arithmetic operations together with an histogram equalization followed by an automatic thresholding and applied to gray images, is proposed in [8]. After applying all the proposed steps, not only a few artifacts persist in the resulting image, but also the cell nucleus are enhanced. More recently, in [9] was proposed a method that accurately segment WBC smear images, after which it was carried out the classification in five classes of cells by means of a naïve Bayesian classifier. In the approach, and for the segmentation step, authors use RGB images, and more specifically by summing the R and B channels, and dividing the result by the G channel; after those operations, they applied some morphological operations, in order to obtain the cell nucleus.

A technique based on Otsu and the use of circular histograms is proposed in [10] for the WBC segmentation problem; in the approach, smear images are processed in the Hue-Saturation-Intensity (HSI) space, as authors consider that most information about the WBC is contained in the hue and saturation components. An improvement to a fuzzy cellular neural network (IFCNN) was proposed in [11], as an advancement in WBC automatic segmentation. Even though the proposal is very important, it has serious limitations, such as its restricted capability to detect a maximum of two WBC per run, and its performance commonly decays when the iteration number is not properly defined, yielding a challenging problem itself with no clear clues on how to make the best choice.

In digital images, WBC can have some resemblance to circular or ellipsoid shapes; therefore, could be used techniques from computer analysis that detect ellipses, or even circles [12]. Since several decades, ellipse detection in real images is considered an open research problem, for which many approaches that belong to at least two categories have been proposed: Hough transform-based (HT) and Random sampling. Generally speaking, in the detection process are considered some edge points contained into an image to model a candidate ellipse, and later this candidate is compared against the remaining edge points in the image. However, to perform a better ellipse detection in digital images, it is common the use of the Hough Transform [13, 14]. This technique works by representing geometric shapes by its parameters, and then accumulating its respective bins in the parameter space. Peaks in the bins provide a visual insight of where ellipses may be; therefore, since the parameters are quantized into discrete bins, the intervals of the bins directly affect the accuracy of the results and the computational effort. As a consequence, finer quantizations are needed to get more accurate results, but to the expense of large memory loads and large computation times. To reduce such problems, several improvements to the original Hough Transform have been proposed for ellipse detection by researchers; an approach is the random sampling, in which a bin represents a candidate shape rather than a set of quantized parameters [15–17]. It has been shown that a random sampling-based approach produces improvements in accuracy and computational complexity, as well as a reduction in the number of false positives (incxistent ellipses), when compared to the original HT and some of its improved variants [18]. Even though this approach reduces the associated problems, these are not eliminated, because it is similar to the HT in the sense that goes through an accumulation process for the bins; accordingly, the bin with the highest score represents the best approximation of an actual ellipse in the target image.

Since several years, another method for ellipse detection is to consider this problem as an optimization problem, which could be solved by using traditional and nontraditional optimization techniques, such as evolutionary algorithms. In general, they have demonstrated to give better results than those based on the HT and random sampling with respect to accuracy and robustness [19]. Such approaches have produced several robust ellipse detectors using different optimization algorithms such as Genetic algorithms (GA) [20, 21] and Particle Swarm Optimization (PSO) [22], among other. The main advantage with respect to traditional methods for ellipse detection, such as HT and random sampling, is that the optimization approach could serve to detect even partially occluded, deformed, or incomplete ellipses, semielliptical forms, etc.; nonetheless, in spite of those benefits, this concept has been scarcely applied to WBC detection in medical images. An interesting exception is the work presented in [23], in which it is used a GA for the WBC detection; however, the authors only considers circles with fixed radius in the fitness function, provoking false positives, particularly for images that contained overlapped or irregular WBC.

This chapter deals with the automatic detection of WBC by using a DE algorithm [24]; this is a novel, population-based, evolutionary algorithm which has been used to optimize complex continuous nonlinear functions; this metaheuristic employs simple mutation and crossover operators to generate candidate solutions, and utilizes a simple competition scheme to decide if either the candidate or the parent will survive to the next generation. Due to its simplicity, ease of implementation, fast convergence, and robustness, this algorithm has been applied to solve several engineering problems, as reported in specialized literature [25–29].

The method uses five points from the edge map to encode candidate ellipses, whereas an objective function measures the similarity between that ellipse and the edge points. Guided by the values of such objective function, the set of encoded candidate ellipses are evolved using the DE algorithm so they can fit into actual WBC on the image. The approach generates a sub-pixel detector which can effectively identify leukocytes in real images, even in those with complex conditions, such as occlusions, missing information, etc. The WBC detector based on DE is compared with other detectors on multiple smear images.

The sections on this chapter are arranged as follows: Sect. 8.2 explains in detail the Differential Evolution algorithm as an optimization method, while Sect. 8.3 clarifies the ellipse detection problem from an optimization perspective. Section 8.4 presents the complete WBC detector, and Sect. 8.5 reports the obtained experimental results obtained by using DE as optimization tool; as a complement, in Sect. 8.5.2 are presented some comparisons with other methods utilized for WBC detection. Section 8.6 concludes this chapter.

8.2 Differential Evolution for Optimization

For the most part, optimization is a searching tool, whose goal is to find the best element from a given set of objects, usually real numbers, and by considering an objective function; frequently, those groups of objects are huge. For that reason, in the majority of the cases an exhaustive search is not an option. Therefore, sometimes the use of evolutionary algorithms to make intelligent quests is a good option. As other evolutionary algorithms for optimization, DE is population based, its operators are simple and efficient, and it is capable to search for the global optimum in complicated functions; accordingly, this algorithm has been used to solve several problems in engineering and research. There exist several versions of DE, which differ in the way the parents as well as their number are selected [24]; however, in this chapter it is used the version called *"rand-to-best/1/bin"* in the experimental part. The first step in the algorithm is the population initialization; that population is composed by N_P vectors of D dimensions, randomly generated from a uniform distribution between the box restrictions $x_{j,\text{high}}$ and $x_{j,\text{low}}$:

$$x_{j,i,t} = x_{j,\text{low}} + \text{rand}(0,1) \cdot (x_{j,\text{high}} - x_{j,\text{low}});$$
$$j = 1, 2, \ldots, D; \quad i = 1, 2, \ldots, N_p; \quad t = 0. \tag{8.1}$$

where the subscript t represents the actual iteration, while i and j are the parameter and particle indexes, respectively. Therefore, $x_{j,i,t}$ is the jth parameter of the ith particle in generation t. At each iteration and for each parent in the population, a trial solution must be generated taking the best solution found to the moment $\mathbf{x}_{best,t}$ in the current population, and adding the weighted difference between two randomly selected parents from the population:

$$\mathbf{v}_{i,t} = \mathbf{x}_{best,t} + F \cdot (\mathbf{x}_{r_1,t} - \mathbf{x}_{r_2,t});$$
$$r_1, r_2 \in \{1, 2, \ldots, N_p\} \tag{8.2}$$

with $\mathbf{v}_{i,t}$ being the trial vector in the current iteration for the parent i, and where r_1 and r_2 are uniform random numbers restricted to $r_1 \neq r_2 \neq i$; the mutation scale factor F is a positive real number, typically less than one. In Fig. 8.1 is illustrated the vector-generation process defined by Eq. (8.2).

Once the mutated vector is obtained, a crossover between this and the current parent $\mathbf{x}_{i,t}$ is performed to acquire the trial vector, which is computed as follows:

$$u_{j,i,t} = \begin{cases} v_{j,i,t}, & \text{if } \mathrm{rand}(0,1) \leq CR \text{ or } j = j_{\mathrm{rand}}, \\ x_{j,i,t}, & \text{otherwise} \end{cases} \tag{8.3}$$

considering that $j_{\mathrm{rand}} \subset \{1, 2, \ldots, D\}$. In order to probabilistically control how many phenotypes will be changed in the crossover operation to retrieve the trial vector, a parameter is tuned, usually with a value defined as $(0.0 \leq CR \leq 1.0)$. Additionally, by using the randomly chosen index j_{rand}, it is assured that at least one parameter will be different from the original parent $\mathbf{x}_{i,t}$.

In the last part of the DE algorithm, it is applied a selection operator, in order to make an elitist choice of the new parent, that will survive to the next generation; this is done in a greedy form: if the computed cost function value of the trial vector $\mathbf{u}_{i,t}$ is better than the cost of the vector $\mathbf{x}_{i,t}$, then such trial vector replaces $\mathbf{x}_{i,t}$ in the next generation. Otherwise, $\mathbf{x}_{i,t}$ remains in the population for at least one more generation

$$\mathbf{x}_{i,t+1} = \begin{cases} \mathbf{u}_{i,t}, & \text{if } f(\mathbf{u}_{i,t}) \leq f(\mathbf{x}_{i,t}), \\ \mathbf{x}_{i,t}, & \text{otherwise} \end{cases} \tag{8.4}$$

Fig. 8.1 Example of a mutated vector generated with the DE/best/1/exp version

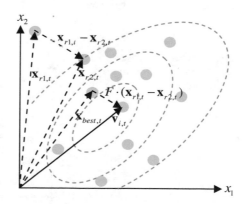

where f represents the objective function. The operators are sequentially applied to the whole population to complete a run (or generation), and it is repeated either until a termination criterion is attained or a predefined generation number is reached.

8.3   Ellipse Detection

8.3.1   Candidate Ellipses

In the methodology presented in this chapter, the first part consider that a gray scale image will be transformed to an corresponding edge map, from which the candidate ellipses will be constructed. Therefore, after a filtering process to enhance borders, it is usually utilized an edge detector, such as the Canny filter. Then, the (x_i, y_i) coordinates for each edge pixel p_i are stored inside the edge vector $P = \{p_1, p_2, \ldots, p_{Ne}\}$, with N_e being the total number of edge pixels contained in an edge map of the image. In order to draw different geometric figures in image processing, there is a minimum requirement of points; for example, lines will demand two points, circles three, etc. In the case of ellipses, five points are necessary to configure a candidate solution, or candidate ellipse, E. Such points are uniformly and randomly selected from the edge array P; this procedure will encode a candidate solution as the ellipse that passes through five points p_1, p_2, p_3, p_4 and p_5 ($E = \{p_1, p_2, p_3, p_4, p_5\}$). Accordingly, by substituting the coordinates of those points into Eq. 8.5, we gather a set of five simultaneous equations with five unknown parameters a, b, f, g and h:

$$ax^2 + 2\,hxy + by^2 + 2\,gx + 2fy + 1 = 0 \tag{8.5}$$

Taking as a starting point the edge points shown in Fig. 8.2, it is possible to calculate the ellipse center (x_0, y_0), the maximum radius $r_{\max}$, the minimum radius $r_{\min}$, and the ellipse orientation (θ), as follows:

Fig. 8.2 Candidate ellipse formed from the points p_1, p_2, p_3, p_4 and p_5

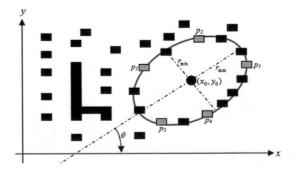

$$x_0 = \frac{hf - bg}{C}, \tag{8.6}$$

$$y_0 = \frac{gh - af}{C}, \tag{8.7}$$

$$r_{max} = \sqrt{\frac{-2\Delta}{C(a+b-R)}}, \tag{8.8}$$

$$r_{min} = \sqrt{\frac{-2\Delta}{C(a+b+R)}}, \tag{8.9}$$

$$\theta = \frac{1}{2}\arctan\left(\frac{2h}{a-b}\right) \tag{8.10}$$

where

$$R^2 = (a-b)^2 + 4h^2, C = ab - h^2 \quad \text{and} \quad \Delta = \det\left(\begin{vmatrix} a & h & g \\ h & b & f \\ g & f & 1 \end{vmatrix}\right). \tag{8.11}$$

8.3.2   Objective Function

In this chapter, to calculate the error produced by a candidate solution E, the ellipse coordinates are calculated as a virtual shape which must be validated, i.e. it is necessary to verify its existence in the edge image. This is achieved by conforming a test set $S = \{s_1, s_2, \ldots, s_{N_s}\}$, where N_s are the number of points over which the existence of an edge point, corresponding to E, should be tested.

The Bresenham algorithm to draw circular arcs is used to generate the set S; this method is called Midpoint Circle Algorithm [30], and seeks the required points for drawing an circle, which of course could be extended to other circular forms such as ellipses [31], in whose case is called Midpoint Ellipse Algorithm (MEA). Any point (x, y) on the boundary of the ellipse with the parameters a, b, f, g and h, must satisfies the equation $f_{ellipse}(x, y) \cong r_{max}x^2 + r_{min}y^2 - r_{max}^2 r_{min}^2$, where r_{max} and r_{min} represent the major and minor axis, respectively. However, MEA avoids computing square-root calculations by comparing the pixel separation distances. A method for direct distance comparison is to test the halfway position between two pixels (sub-pixel distance) to determine if this midpoint is inside or outside the ellipse boundary. If the point is in the interior of the ellipse, the ellipse function is negative. Thus, if the point is outside the ellipse, the ellipse function is positive. Therefore, the error involved in locating pixel positions using the midpoint test is limited to one-half the pixel separation (sub-pixel precision). To summarize, the relative position of any point (x, y) can be determined by checking the sign of the ellipse function:

$$f_{ellipse}(x, y) \begin{cases} < 0 & \text{if } (x, y) \text{ is inside the ellipse boundary} \\ = 0 & \text{if } (x, y) \text{ is on the ellipse boundary} \\ > 0 & \text{if } (x, y) \text{ is outside the ellipse boundary} \end{cases} \qquad (8.12)$$

This function is applied to every mid-positions between pixels nearby the ellipse path at each sampling step. Figure 8.3a, b show the midpoint evaluation in two different positions of a candidate ellipse. In order to draw the ellipse, the virtual space is divided in octants, as depicted in Fig. 8.4.

The computation time used by MEA is substantially reduced, as the algorithm pays especial attention to the symmetry of the ellipses; ellipse sections in adjacent octants within one quadrant are symmetric with respect to the line dividing the two octants, as shown in Fig. 8.4. The algorithm can be considered as the quickest, providing at the same time a sub-pixel precision. However, in order to protect the MEA operation, it is important to assure that points lying outside the image plane must not be considered in S. After 'drawing' the candidate ellipse by using MEA, it is necessary to validate its presence in the edge map; the objective function used to measure the matching error between the candidate pixel coordinates from S and the edge map is:

$$J(E) = 1 - \frac{\sum_{v=1}^{Ns} G(x_v, y_v)}{Ns} \qquad (8.13)$$

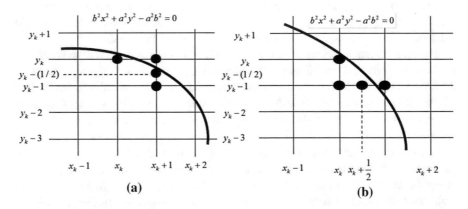

(a)   **(b)**

Fig. 8.3 **a** Symmetry of the ellipse: an estimated one octant which belong to the first region where the slope is greater than -1, **b** In this region the slope will be less than -1 to complete the octant and continue to calculate the same so the remaining octants

Fig. 8.4 Midpoint between candidate pixels at sampling position x_k along an elliptical path

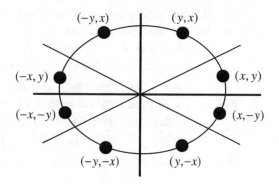

considering that $G(x_i, y_i)$ is a function which corroborates whether the virtual pixel with coordinates $(x_v, y_v) \in S$ has a corresponding edge point with coordinates (x, y) in the edge image:

$$G(x_v, y_v) = \begin{cases} 1 & \text{if the pixel } (x_v, y_v) \text{ is an edge point} \\ 0 & \text{otherwise} \end{cases} \tag{8.14}$$

A value of $J(E)$ approaching to zero means that a good candidate ellipse was found by the algorithm, and a zero value means a 100% correspondence between the virtual ellipse and the edge points in the image. Let's consider the procedure to evaluate a candidate, as shown in Fig. 8.5; from the edge map example (Fig. 8.5a) are taken five points $E = \{p_1, p_2, p_3, p_4, p_5\}$, which are used to make a candidate ellipse S (Fig. 8.5b). After that, the virtual ellipse is compared point to point against the edge points in the image; by following the example, in this case the virtual shape S has 52 points ($N_s = 52$), from which 35 of them in the edge map (Fig. 8.5c). Therefore, $\sum_{v-1}^{N_s} G(x_v, y_v) = 35$, and the objective function value is J $(E) = 0.327$.

8.3.3 Implementation of DE for Ellipse Detection

The ellipse detector algorithm based on DE can be summarized in the following steps:

Step 1:   Set the DE parameters F=0.25 and CR=0.8.

Step 2:   Initialize the population of m individuals $\mathbf{E}^k = \{E_1^k, E_2^k, ..., E_m^k\}$
where each decision variable p_1, p_2, p_3, p_4 and p_5 of E_a^k is set randomly within the interval $[1, N_p]$. All values must be integers. Considering that k=0 and $a \in (1, 2, ..., m)$.

Step 3:   Evaluate the objective value $J(E_a^k)$ for all m individuals, and determining the $E^{best,k}$ showing the best fitness value, such that
$E^{best,k} \in \{\mathbf{E}^k\} \big| J(E^{best,k}) = \min\{J(E_1^k), J(E_2^k), ..., J(E_m^k)\}$.

Step 4:   Generate the trial population $\mathbf{T} = \{T_1, T_2, ..., T_m\}$:

> for (i=1; i<m+1; i++)
>   do r_1 =floor(rand(0,1)$\cdot m$); while ($r_1 = i$);
>   do r_2 =floor(rand(0,1)$\cdot m$); while (($r_2 = i$) or ($r_2 = r_2$));
>   jrand=floor(5 $\cdot$ rand(0,1));
>   for (j=1; j<6; j++) // generate a trial vector
>   if (rand(0,1)<=CR or j=jrand)
> $T_{j,i} = E_j^{best,k} + F \cdot (E_{j,r_1}^k - E_{j,r_2}^k)$;
>
>   else
>
> $T_{j,i} = E_{j,i}^k$;
>
>   end if
>   end for
> end for

Step 5:   Evaluate the fitness values $J(T_i)$ ($i \in \{1, 2, ..., m\}$) of all trial individuals. Check all individuals. If a candidate parameter set is not physically plausible, i.e. out of the range $[1, N_p]$, then an exaggerated cost function value is returned. This aims to eliminate "unstable" individuals.

Step 6:   Select the next population $\mathbf{E}^{k+1} = \{E_1^{k+1}, E_2^{k+1}, ..., E_m^{k+1}\}$:

> for (i=1; i<m+1; i++)
>   if ($J(T_i) < J(E_i^k)$)
> $E_i^{k+1} = T_i$
>   else
> $E_i^{k+1} = E_i^k$
>   end if
> end for

Step 7:   If the iteration number (NI) is met, then the output $E^{best,k}$ is the solution (an actual ellipse contained in the image), otherwise go back to Step 3.

Fig. 8.5 Five points are used to draw a virtual ellipse, which is compared against the edge points

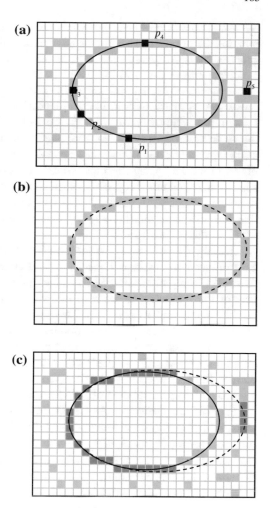

8.4 Detecting White Blood Cells

Several steps are needed to correctly detect the WBC in digital images; they integrate a thresholding technique with an ellipse detection, such as the presented in the previous section. In next paragraphs are presented the necessary steps to configure a WBC detector together with an example.

8.4.1 Segmentation and Border Extraction

In this step it is assumed that the smear images have been previously digitized; then, they must be preprocessed in order to get segmented as well as edge map images, by using both a thresholding technique, and a border extraction algorithm. As mentioned in previous section, the edge map is necessary to evaluate each potential solution by the objective function, in order to measure the resemblance of that candidate ellipse with a WBC really existing in the image.

Segmentation techniques are utilized as a preprocessing step in many visual processing systems, both to isolate objects of interest, and to reduce the memory burden. Accordingly, in this case the goal of the segmentation strategy is to isolate the white blood cells (WBC's) from other structures such as red blood cells as well as background pixels. There exists several thresholding schemes used to classify the pixels in images, being the most common those which utilize image information such as color, texture, brightness, or gradient; however, due to the characteristics of the image's family considered in this chapter, it could be used a simpler technique that use histogram thresholding to segment WBC, like the Diffused Expectation-Maximization (DEM) [32].

This algorithm is based on the classical Expectation-Maximization, and it has been utilized to segment images, in particular medical ones [33]. Different to classical EM algorithms, DEM considers the spatial correlations among pixels as a part of the minimization criteria, allowing object segmentation regardless noisy and/or other complex conditions. The method models an image as a finite mixture, considering that each mixture component corresponds to a region class, and uses a maximum likelihood approach to estimate the parameters of each class, via the expectation maximization (EM) algorithm, coupled with anisotropic diffusion on classes, in order to account for the spatial dependencies among pixels.

A known implementation of DEM, contributed by [34], was used in this chapter for the WBC segmentation. As this DEM version has the capacity to segment both color as well as gray-level images, it can be used for smear digital images regardless the way they are acquired. After several manual tests, the DEM was tuned for three classes ($K = 3$) with the parameters: $g(\nabla h_{ik}) = |\nabla h_{ik}|^{-9/5}$, $\lambda = 0.1$ and $m = 10$ iterations; those values are in agreement with [32].

Three segmentation points are obtained after applying the DEM: one corresponding to the WBCs, other to red blood cells, and the last one which pertains to the background. In Fig. 8.6 are shown the original smear's image, the segmentation of WBC, as well as the edge map.

As can be seen in the preprocessed image, the last step is the edge map computation. The principal aim of the edge map transformation is to retain the object structures, whereas the memory burden is reduced; by considering that, the DE-based detector operates directly over the edge map in order to recognize ellipsoidal shapes. Even though have been proposed several algorithms to perform border extraction, and for the sake of simplicity, it is that in this chapter it is used the morphological edge detection procedure [35]. This is a simple method to extract borders from binary images in which original images (I_B) are eroded by a minimum structure element (I_E) composed by a template of 3×3 ones. After the previous operation, the resulting image is inverted ($\bar{I}_E$) and compared with the original ($\bar{I}_E \wedge I_B$); the result is the edge

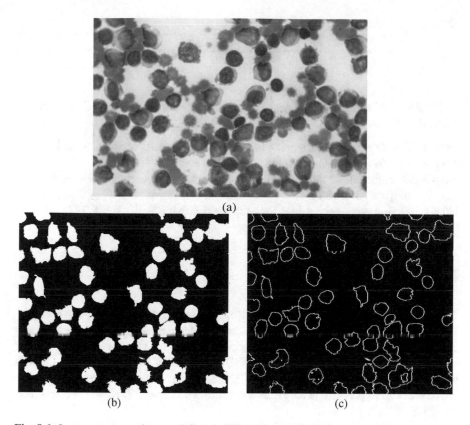

Fig. 8.6 Image pre-processing: **a** original, **b** WBC segmentation, **c** edge map

map of I_B. An example of the application of those morphological operations is shown in Fig. 8.6c.

8.4.2   Ellipse Approximation of WBC

The object structures (e.g. edge map), obtained from the methodology shown in previous section, are used as input data for the ellipse detector presented in Sect. 8.3, and the parameter configuration used in this chapter for the DE algorithm are shown in Table 8.1. It is important to notice that the final configuration coincides with a good possible calibration proposed in [36], where it has been analyzed the effect of modifying the DE-parameters in several generic optimization problems. Also, it was chosen a population size ($m = 20$) that provides a good balance between a fast convergence and the computational overload. After those algorithm calibrations are completed, they are kept for all the experimental part in the remaining of this chapter.

Table 8.1 DE tuning parameters

m	F	CR	NI
20	0.25	0.80	200

Finally, the approach displayed in this chapter for WBC detection is implemented as follows:

Step 1   Segment the WBC's using the DEM algorithm (described in Sect. 8.4.1).
Step 2   Get the edge map from the segmented image.
Step 3   Start the ellipse detector based in DE over the edge map while saving best ellipses (Sect. 8.3).
Step 4   Define parameter values for each ellipse that identify the WBC's.

8.4.3   A Simple Example

Prior to the experimental part, in this section it is depicted the algorithm in a step by step fashion through a numeric example that has been set by applying the proposed method to detect a single leukocyte lying inside of a simple image (Fig. 8.8a). After applying the threshold operation, the WBC is located besides few other pixels which are merely noise (see Fig. 8.7b). Then, the edge map is subsequently computed and stored pixel by pixel inside the vector P. Figure 8.7c shows the resulting image after such procedure.

The DE-based ellipse detector is executed using information from the edge map (for the sake of easiness, it is only considered a population of four particles). Like all evolutionary approaches, DE is a population-based optimizer that attacks the starting point problem by sampling the search space at multiple, randomly chosen, initial particles. By taking five random pixels from vector P, four different particles are constructed. Figure 8.7d depicts the initial particle distribution $\mathbf{E}^0 = \{E_1^0, E_2^0, E_3^0, E_4^0\}$. By using the DE operators, four different trial particles $\mathbf{T} = \{T_1, T_2, T_3, T_4\}$ (ellipses) are generated, whose locations are shown in Fig. 8.7e. Then, the new population $\mathbf{E}^1$ is selected considering the best elements obtained among the trial elements $\mathbf{T}$ and the initial particles $\mathbf{E}^0$. The final distribution of the new population is depicted in Fig. 8.7f. Since the particles E_2^0 and E_2^0 hold (in Fig. 8.7f) a better fitness value ($J(E_2^0)$ and $J(E_3^0)$) than the trial elements T_2 and T_3, they are considered as particles of the final population $\mathbf{E}^1$. Figure 8.8g, h present the second iteration produced by the algorithm whereas Fig. 8.7i shows the population configuration after 25 iterations. From Fig. 8.7i, it is clear that all particles have converged to a final position which is able to accurately cover the WBC. Also, as the problem of WBC detection is considered as an optimization problem, it is understandable that at the last iteration the final population of candidate ellipses will not share the same fitness value, but slight different values.

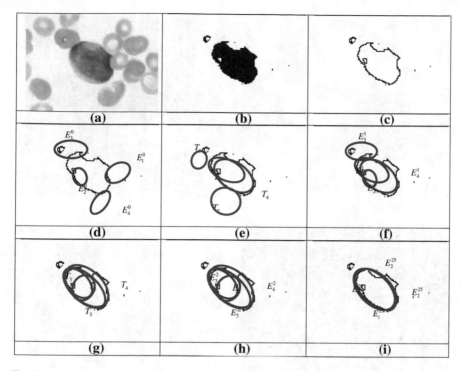

Fig. 8.7 A simple detection example

8.5 Experimental Results

8.5.1 Simple Experiments

In the experimental part, several tests have been proposed in order to evaluate the performance of the WBC detector. The utilized images belong to microscope images taken from blood-smears with a 960×720 pixel resolution, and they correspond to several samples used to support a leukemia diagnosis. This family of images present several and different complex conditions such as deformed cells, overlapping, partial occlusion, or even combinations of them. In that sense, the robustness of the algorithm has been tested under such demanding conditions. All the experiments has been developed using a PC based on Intel Core i7-2600, with 8 GB in Ram.

In Fig. 8.8a it is displayed an example of the kind of images employed in this experimental part; those images were used as input for the WBC detector; Fig. 8.8b presents the segmented WBC's obtained by the DEM algorithm, whereas the edge map and the white blood cells after the detection process, are respectively shown in Fig. 8.8c, d. The results show that this algorithmic approach can effectively detect and mark blood cells despite cell occlusion, deformation or overlapping. Moreover, once

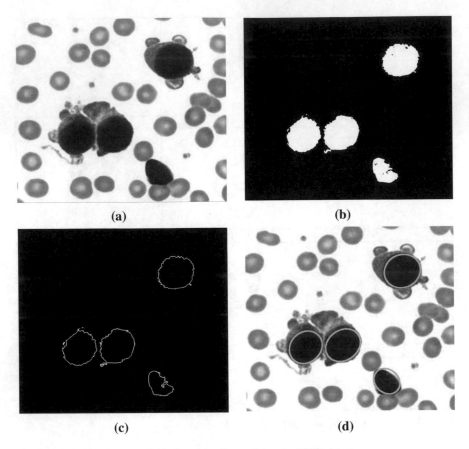

Fig. 8.8 Resulting images of the first test after applying the WBC detector

the WBC have been detected, then other parameters could be calculated, such as the total area covered by the blood cells, or even relationships between several cell sizes.

Other example is presented in Fig. 8.9, where the image shows seriously deformed cells, and therefore represents a complex example. Despite such imperfections, the WBC detector can effectively detect the cells as it is shown in Fig. 8.9d, even the partial parts of the cells at the bottom of the image.

8.5.2 Comparing WBC Detectors Over Simple Images

A comprehensive set of smear-blood test images was used to test the performance of the WBC detector. We have applied this DE-based detector to test images in order to compare its performance against other WBC detection algorithms such as the Boundary Support Vectors (BSV) approach [7], the iterative Otsu (IO) method [10], the Wang algorithm [11] and the Genetic algorithm-based (GAB) detector

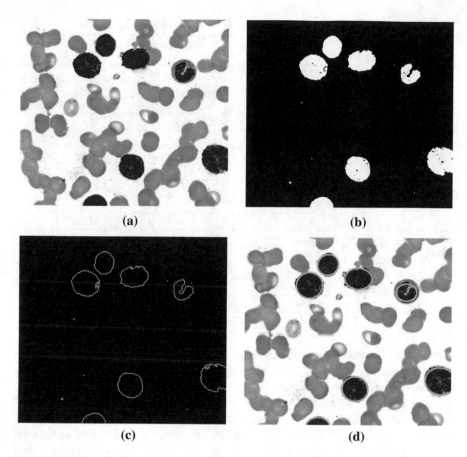

(a)   (b)

(c)   (d)

Fig. 8.9 Resulting images of the second test after applying the WBC detector

[23]. In all cases, the algorithms were tuned according to the value set which was originally proposed by their own references.

An issue also considered in the experimental part was the detection performance of the methods used in this chapter; in that sense, Table 8.2 tabulates the comparative leukocyte detection performance of the BSV approach, the IO method, the Wang algorithm, the BGA detector and the DE-based method, in terms of detection rates and false positives.

The images used in the experimental part were taken from the ASH Image Bank (http://imagebank.hematology.org/), and compose a set of 50 images. In total, they have 517 leukocytes (287 bright leukocytes and 230 dark leukocytes according to smear conditions), according to a human expert counting. Such images represents the ground truth used in this chapter as a reference for all the experiments. Two indexes were used for the sake of comparisons: the detection rate (DR), which is defined as the ratio between the number of leukocytes correctly detected and the number leukocytes determined by the expert, and false positives rate (FPR),

Table 8.2 Comparison among the used method for WBC detection

Leukocyte type	Method	Leukocytes detected	Missing	False alarms	DR (%)	FPR (%)
Bright (287)	BSV	130	157	84	45.30	29.27
	IO	227	60	73	79.09	25.43
	Wang	231	56	60	80.49	20.90
	BGA	220	67	22	76.65	7.66
	DE-based	281	6	11	97.91	3.83
Dark (230)	BSV	105	125	59	46.65	25.65
	IO	183	47	61	79.56	26.52
	Wang	196	34	47	85.22	20.43
	BGA	179	51	23	77.83	10.00
	DE-based	227	3	3	98.70	1.30
Overall (517)	BSV	235	282	143	45.45	27.66
	IO	410	107	134	79.30	25.92
	Wang	427	90	107	82.59	20.70
	BGA	399	118	45	77.18	8.70
	DE-based	508	9	14	98.26	2.71

described as he ratio between the number of non-leukocyte objects that have been wrongly identified as leukocytes and the number leukocytes which have been actually determined by the expert.

As can be seen from Table 8.2, after run the comparison experiments, the WBC detector based on DE gives the best results, because it achieves a 98.26% accuracy in leukocyte detection with a 2.71% false positive rate, when compared against the remaining algorithms; only as a reference, the second best algorithm is more than three times worst than the DE-based method.

8.5.3   Comparing WBC Detectors Over Complex Images

As in any other digital image processing, one of the main error sources is the addition of noise to the original image; accordingly, images of blood smear are often deteriorated by noise due to various sources of interference, such as a bad smear preparation, poor illumination, camera problems, etc. It is therefore a desirable characteristic that the detection method could cope with such problems, because the process' success highly depends on its ability to overcome and cope with different kind of noises. In this section, all the algorithms are tested and compared for the sake of robustness when the experimental images are contaminated by noise. Thus, two experiments were proposed. On the one hand, it is examined the performance of each algorithm when the detection task is accomplished over images corrupted by Salt & Pepper noise. On the other hand, it is investigated the algorithm's behaviour when the images were contaminated by Gaussian noise. These kind of noise represent the most usually encountered in images of blood smear. As in the previous section, in this one it is considered the set of 50 smear blood images, which contain 517 leukocytes which have been detected and counted by a human expert. The added noise is produced by MatLab, considering two noise levels of 5% and 10% for Salt & Pepper noise whereas $\sigma = 5$ and $\sigma = 10$ are used for the case of Gaussian noise. Such noise levels, according to [37], correspond to the best trade of between detection difficulty and real existence in medical imaging. Using higher noise levels, the detection process would be unnecessarily complicated without representing a feasible image condition.

In Fig. 8.10 are shown two examples of the experimental images that contains added noise, whereas in Tables 8.3 and 8.4 are reported the outcomes of the experiments in terms of the detection rate (DR) and false positive rate (FPR).

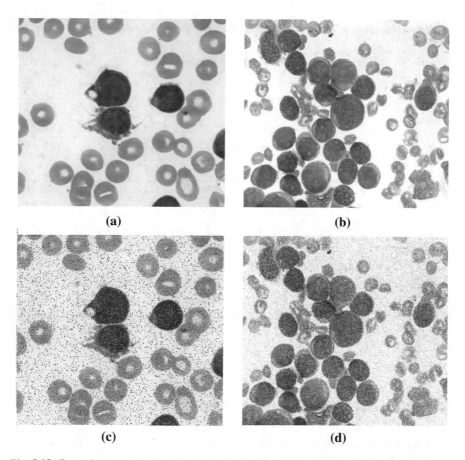

(a)   (b)

(c)   (d)

Fig. 8.10 Some images included in the experimental set: **a–b** Originals images. **c** Image contaminated with 10% of Salt & Pepper noise, and **d** image polluted with $\sigma = 10$ of Gaussian noise

Table 8.3 Comparative WBC detection over images corrupted by different levels of Salt & Pepper noise

Noise level	Method	Leukocytes detected	Missing	False alarms	DR (%)	FPR (%)
5% 517 Leukocytes	BSV	185	332	133	34.74	26.76
	IO	311	206	106	63.38	24.88
	Wang	250	176	121	58.68	27.70
	GAB	298	219	135	71.83	24.18
	DE-based	482	35	32	91.55	7.04
10% 517 Leukocytes	BSV	105	412	157	20.31	30.37
	IO	276	241	110	53.38	21.28
	Wang	214	303	168	41.39	32.49
	GAB	337	180	98	65.18	18.95
	DE-based	463	54	31	89.55	5.99

Table 8.4 Comparative WBC detection over images corrupted by different levels of Gaussian noise

Noise level	Method	Leukocytes detected	Missing	False alarms	DR (%)	FPR (%)
$\sigma = 5$ 517 Leukocytes	BSV	214	303	98	41.39	18.95
	IO	366	151	87	70.79	16.83
	Wang	358	159	84	69.25	16.25
	GAB	407	110	76	78.72	14.70
	DE-based	487	30	21	94.20	4.06
$\sigma = 10$ 517 Leukocytes	BSV	162	355	129	31.33	24.95
	IO	331	186	112	64.02	21.66
	Wang	315	202	124	60.93	23.98
	GAB	363	154	113	70.21	21.86
	DE-based	471	46	35	91.10	6.77

In the results it is clear that the WBC based on DE gives the best detection performance, achieving in the worst case a DR of 89.55 and 91.10%, under contaminated conditions of Salt & Pepper and Gaussian noise, respectively. Also, this approach possesses the least degradation performance presenting a FPR value of 5.99 and 6.77%.

8.5.4 Comparing Stability of WBC Detectors

An important issue for WBC detection is the method's capability to detect in a consistent way the cells present in the image. In that sense, in this section are compared two methods: on one hand, the DE-based method, and on the other hand, the approach proposed by Wang et al. [11], which is considered an accurate technique for WBC detection. The Wang algorithm is an energy minimizing method that is guided by internal constraint elements and influenced by external forces, producing the segmentation results over closed contours. As external forces, the Wang approach uses edge information which is usually represented by the gradient magnitude of the image. Therefore, the contour is attracted to pixels with large image gradients, i.e. strong edges. At each iteration, the Wang method finds a new contour configuration which minimizes the energy that corresponds to external forces and constraint elements. The neural network structure and its operational parameters for the comparison were taken from [11], and the parameters for the DE-based detector were taken from Table 8.1. A result obtained after using the two methods is shown in Fig. 8.11, by using a simple image with two WBC. As can be

Fig. 8.11 Comparing the accuracy of two methods for white blood cell detection: a Original, b Wang's method, c DE method

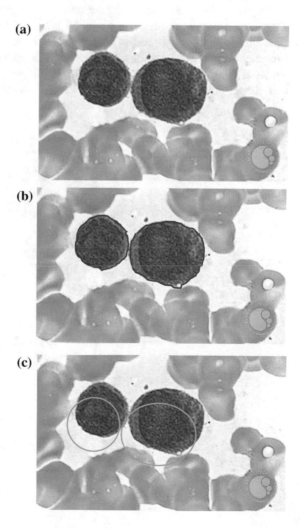

seen, the Wang's method is more accurate if the number of iterations is adequately given (Fig. 8.11b); in this case, both algorithms were run 200 iterations, and the results are pretty similar (Fig. 8.11c).

The Wang algorithm uses the fuzzy cellular neural network (FCNN) as optimization approach. It employs gradient information and internal states in order to find a better contour configuration. In each iteration, the FCNN tries, as contour points, different new pixel positions which must be located nearby the original contour position. Such fact might cause the contour solution to remain trapped into a local minimum. In order to avoid such a problem, the Wang method applies a considerable number of iterations so that a near optimal contour configuration can be found. However, when the number of iterations increases the possibility to cover other structures increases too. Thus, if the image has a complex background (just as smear images do) or the WBC's are too close, the method gets confused so that finding the correct contour configuration from the gradient magnitude is not easy. Therefore, a drawback of Wang's method is related to its optimal iteration number (instability). Such number must be determined experimentally as it depends on the image context and its complexity. Figure 8.12a shows the result of applying 400 cycles of the Wang's algorithm while Fig. 8.12b presents the detection of the same cell shapes after 1000 iterations using the DE-based algorithm. From Fig. 8.12a,

Fig. 8.12 Accuracy of methods for WBC detection: **a** Wang's algorithm after 400 cycles, and **b** DE detector method considering 1000 cycles

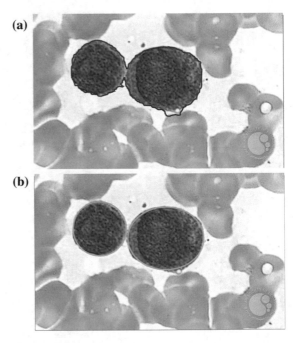

Table 8.5 Error in cell's size estimation after applying the DE algorithm and the Wang's method

Algorithm	Iterations	Error (%)
Wang	30	88
	60	70
	200	1
	400	121
	600	157
DE-based	30	24.30
	60	7.17
	200	2.25
	400	2.25
	600	2.25

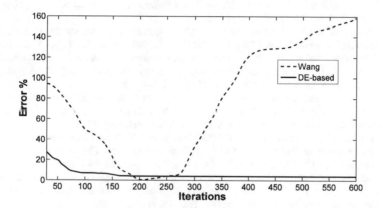

Fig. 8.13 Error % versus iterations of the algorithms after one run

it can be seen that the contour produced by Wang's algorithm degenerates as the iteration process continues, wrongly covering other shapes lying nearby.

The methods used in this sections were compared by using the estimated WBC area obtained by each one; such areas were obtained considering different number of iterations for each algorithm. In Table 8.5 are shown the areas obtained by each method when the iteration's number is changed, and all the experiments were run 20 times for statistical validation. Moreover, the error evolution after one run of the two algorithms is depicted in Fig. 8.13.

8.6 Conclusions

In this chapter were compared five algorithms for WBC detection in digital smear blood images. As the comparison results suggests, the DE-based algorithm is the best option among them for WBC detection, even when the images have the most common noise levels, as was shown in Sect. 8.5. This algorithm considers the complete process as a multiple ellipse detection problem, and uses the encoding of five edge points as candidate ellipses in the edge map of the smear, whereas an objective function allows to accurately measure the resemblance of a candidate ellipse with an actual WBC on the image. Also, it was shown that the DE-based algorithm has the best accuracy for the detection, when compared against the Wang method.

References

1. L., Long, L.R., Antani, S., Thoma, G.R. He, "Histology image analysis for carcinoma detection and grading," *Computer Methods and Programs in Biomedicine, 107 (3), pp. 538–556*, vol. 107, no. 3, pp. 538–556, 2012.
2. Til Aach, Thomas M. Deserno, Torsten Kuhlen Ingrid Scholl, "Challenges of medical image processing," *Comput Sci Res Dev, 26, (2011), 5–13*, vol. 26, no. 2011, pp. 5–13, 2011.
3. L. Burtseva, "Why should we use the non-existent? Advantages of application of unconventional computing to processing of noisy medical images," *E-Health and Bioengineering Conference (EHB)*, vol. 2015, pp. 1–4., 2015.
4. Sajeena T A and Jereesh A S, "Automated cervical cancer detection through RGVF segmentation and SVM classification," *2015 International Conference on Computing and Network Communications (CoCoNet)*, vol. 2015, pp. 663–669, 2015.
5. H. T. G. and V. V. Nair S. U. Abdulla, "A General Approach for Color Feature Extraction of Microorganisms in Urine Smear Images," *Fifth International Conference on Advances in Computing and Communications (ICACC)*, vol. 2015, pp. 338–341, 2015.
6. Andrea Loddo, Lorenzo Putzu Cecilia Di Ruberto, "A Multiple Classifier Learning by Sampling System for White Blood Cells Segmentation," in *Computer Analysis of Images and Patterns.*: Springer International Publishing, 2015, vol. 9257, pp. 415–425.
7. M. Wang and R. Chu, "A novel white blood cell detection method based on boundary support vectors," *Systems, Man and Cybernetics, 2009. SMC 2009. IEEE International Conference on*, vol. 2009, no. 1, pp. 2595–2598, 2009.
8. S.A. Kareem, H. Ariffin, A.A. Zaidan, H.O. Alanazi and B.B. Zaidan H.T. Madhloom, "An Automated White Blood Cell Nucleus Localization and Segmentation using Image Arithmetic and Automatic Threshold," *Journal of Applied Sciences*, vol. 10, no. 1, pp. 959–966, 2010.
9. Jaroonrut, and Pluempitiwiriyawej, Charnchai Prinyakupt, "Segmentation of white blood cells and comparison of cell morphology by linear and naïve Bayes classifiers," *BioMedical Engineering OnLine*, vol. 14, no. 1, pp. 1–19, 2015.
10. P. Zeng, Y. Zhou and C. Olivier J. Wu, "A novel color image segmentation method and its application to white blood cell image analysis," *8th international Conference on Signal Processing*, pp. 1–4, 2006.
11. F.L. Korris, D. Fu S. Wang, "Applying the improved fuzzy cellular neural network IFCNN to white blood cell detection," *Neurocomputing*, vol. 70, no. 2007, pp. 1348–1359, 2007.
12. NoName, "Circle detection on images based on an evolutionary algorithm that reduces the number of function evaluations," vol. 100, pp. 139–167, 2016.
13. M. Nixon H. Muammar, "Approaches to extending the Hough transform," *Proc. Int. Conf. on Acoustics, Speech and Signal Processing ICASSP-89*, vol. 3, no. 1989, pp. 1556–1559, 1989.
14. D. Kerbyson T. Atherton, "Using phase to represent radius in the coherent circle Hough transform," *IEE Colloquium on the Hough Transform, IEEE*, vol. 1993, pp. 1–4.

15. R. Bolles, M. Fischer, "Random sample consensus: A paradigm to model fitting with applications to image analysis and automated cartography," *CACM*, vol. 24, no. 6, pp. 381–395, 1981.
16. O. Yaron, N. Kiryati D. Shaked, "Deriving stopping rules for the probabilistic Hough transform by sequential analysis," *Comput. Vis. Image. Und.*, vol. 63, no. 1996, pp. 512–526, 1996.
17. E. Oja, P. Kultanen L. Xu, "A new curve detection method: Randomized Hough transform (RHT)," *Pattern Recogn. Lett.*, vol. 11, no. 5, pp. 331–338, 1990.
18. L. Koczy J. Han, "Fuzzy Hough transform," *Proc. 2nd Int. Conf. on Fuzzy Systems*, vol. 2, no. 1993, pp. 803–808, 1993.
19. V., Garcia-Capulin, C. H., Perez-Garcia, A. and Sanchez-Yanez, R. E. Ayala-Ramirez, "Circle detection on images using genetic algorithms," *Pattern Recognition Letters*, vol. 27, no. 2006, pp. 652–657, 2006.
20. Martinez P Lutton E, "A genetic algorithm for the detection of 2D geometric primitives in images," *Proceedings of the 12th international conference on pattern recognition*, vol. 1, no. 1994, pp. 526–528, October 1994.
21. Nawwaf Kharma, Peter Grogono Jie Yao, "A multi-population genetic algorithm for robust and fast ellipse detection," *Pattern Anal Applic*, vol. 8, no. 2005, pp. 149–162, 2005.
22. Yanhui Guo, Yingtao Zhang H.D. Cheng, "A novel Hough transform based on eliminating particle swarm optimization and its applications," *Pattern Recognition*, vol. 42, no. 9, pp. 1959–1969.
23. M. Rangoussi G. Karkavitsas, "Object localization in medical images using genetic algorithms," *World Academy of Science, Eng. and Tec.*, vol. 2, no. 2005, pp. 6–9, 2005.
24. Price K. Storn R, "Differential evolution - a simple and efficient adaptive scheme for global optimization over continuous spaces," International Computer Science Institute, Berkley, Technical Rep. TR-95-012, 1995.
25. Munawar S. Babu B, "Differential evolution strategies for optimal design of shell-and-tube heat exchangers," *Chem Eng Sci.*, vol. 62, no. 14, pp. 3720–3739, 2007.
26. Kinghorn B, Archer A. Mayer D, "Differential evolution – an easy and efficient evolutionary algorithm for model optimization," *Agr Syst*, vol. 83, no. 2005, pp. 315–328, 2005.
27. Mary Raja Slochanal S, Padhy N. Kannan S, "Application and comparison of metaheuristic techniques to generation expansion planning problem," *IEEE Trans Power Syst*, vol. 20, no. 1, pp. 466–475, 2003.
28. Chang C, Su C. Chiou J, "Variable scaling hybrid differential evolution for solving network reconfiguration of distribution systems," *IEEE Trans Power Syst*, vol. 20, no. 2, pp. 668–674, 2005.
29. D. Zaldivar, M. Pérez-Cisneros E. Cuevas, "A novel multi-threshold segmentation approach based on differential evolution optimization," *Expert Systems with Applications*, vol. 37, no. 2010, pp. 5265–5271, 2010.
30. J.E. Bresenham, "A Linear Algorithm for Incremental Digital Display of Circular Arcs," *Communications of the ACM*, vol. 20, no. 1977, pp. 100–106, 1977.
31. J R. Van Aken, "Efficient ellipse-drawing algorithm," *IEEE Comp, Graphics applic.*, vol. 4, no. 9, pp. 24–35, 2005.
32. M. Ferraro, P. Napoletano G. Boccignone, "Diffused expectation maximisation for image segmentation," *Electron Letters*, vol. 40, pp. 1107–1108, 2004.
33. P. Napoletano, V. Caggiano, M. Ferraro G. Boccignonea, "A multi-resolution diffused expectation-maximization algorithm for medical image segmentation," *Computers in Biology and Medicine*, vol. 37, no. 2007, pp. 83–96, 2007.
34. (2012) DEM: Diffused expectation maximization function for image segmentation. [Online]. http://www.mathworks.com/matlabcentral/fileexchange/37197-dem-diffused-expectation-maximisation-for-image-segmentation
35. R.C. Gonzalez and R.E.Woods, *Digital Image Processing*, 1992nd ed., MA Reading, Ed.: Addison Wesley, 1992.

36. F. Huang L. Wang, "Parameter analysis based on stochastic model for differential evolution algorithm," *Applied Mathematics and Computation*, vol. 217, no. 7, pp. 3263–3273, 2010.
37. R. Wagner M. Tapiovaara, "SNR and noise measurements for medical imaging: I. A practical approach based on statistical decision theory," *Physics in Medicine and Biology*, vol. 38, no. 1, pp. 71–92, 1993.

Chapter 9
Estimation of View Transformations in Images

Abstract Computer vision process frequently include an modeling step whose parameters, obtained from a data set, are not easy to calculate due mainly to the presence of a high proportion of outliers. The most known method to overcome this problem is the random sampling consensus (RANSAC). Such technique, in combination with Harmony Search (HS), are used in this chapter for a robust estimation of multiple view relations from point correspondences in digital images. By using this evolutionary technique, the estimation method endorse a different sampling strategy to generate putative solutions: on one hand, RANSAC generate new candidate solutions in a random fashion; on the other hand, by using HS in combination with RANSAC, each new candidate solution is generated by considering the quality of previous candidate solutions. In other words, the solutions space is searched in an intelligent manner. The HS algorithm is inspired by the improvisation process of an orchestra that takes place when musicians search for a better state of harmony; as a result, the HS-RANSAC can substantially reduce the number of iterations still preserving the robust capabilities of RANSAC. The method is used in this chapter to solve the estimation of homographies, with an engineering application to solve the problem of position estimation in a humanoid robot. In this chapter, seven techniques are compared for the problem of robust homography estimation.

9.1 Introduction

In image processing, it is sometimes necessary to take several images from the same scene under different perspectives, and later in the process it could be necessary to estimate geometric relations in order to find an adequate global transformation which superimpose all these images. The diverse scenes could be taken considering a single static camera and a moving object, or even a static object in front of multiple cameras. This methodology has many applications, as for instance, to join together several single images in one panoramic image [1–3]. In order to enhance

E. Cuevas et al., *Evolutionary Computation Techniques:*
A Comparative Perspective, Studies in Computational Intelligence 686,
DOI 10.1007/978-3-319-51109-2_9

the overlapped region, several super-resolution approaches could be utilized, such as the proposed in [4–6]. Other uses of the geometric relations include: the motion estimation of a moving object [7], the calibration of a distributed camera network [8–10], and to control or estimate a robot's position based on the fundamental matrix [11–13]. In a modelling problem, the data that fits into a given hypothetical model are called inliers, whereas other points, e.g., those generated by matching errors, are called *outliers*; they are caused by non-controlled external effects, not related to the investigated model. For the robust classification of those points (e.g. *inliers*, *outliers*), have been proposed some techniques that operates under different criteria; a well known method is the random sampling consensus algorithm, RANSAC [14–17]. This method considers a simple hypothesize-and-evaluation process, under which a minimum set of elements, also called correspondences, is randomly sampled, and a candidate model is determined by using such set. After that, the candidate model is evaluated on the entire dataset separating all their elements into inliers and outliers, according to their degree of matching to such candidate model. Those steps are repeated until the candidate model with a high accuracy is found; usually, the model with the largest number of inliers is considered as the best estimation result. Even thought that this algorithm is simple and straightforward, and that it has been used to solve several problems in engineering, it is clear that also presents some shortcoming, being the two most important [18, 19]: the high consumption of iterations, and the inflexible definition of its objective function. The RANSAC technique generates the candidate models trough a random selection of data samples. Since such a strategy is completely random, then a large number of iterations could be required to explore a representative subset of noisy data and to find a reliable model that could contain the maximum number of inliers. However, the number of iterations increases as a direct consequence of the contamination level of the dataset. Another important issue is the objective function used to evaluate the accuracy of the candidate model over the data, particularly the contaminated data. As the RANSAC methodology considers that the best model is the one that maximizes the number of inliers, then the objective function involves the counting, one by one, of the number of inliers associated to the candidate model; such objective function is susceptible to obtain suboptimal solutions under different circumstances [19]. In order to overcome those two important shortcomings, several variants to the original RANSAC method have been proposed, such as the MLESAC [20], which searches the best hypothesis by maximizing the likelihood via the RANSAC process and assuming that the inlier data would distribute as a Gaussian function whereas outliers are distributed randomly. A different focus was given in the SIMFIT technique [21], in which instead of giving *a-priori* the error scale, it is proposed its prediction by an iterative mechanism. Other important works that improve the original RANSAC technique are the projection-pursuit method [22] and the TSSE (Two-Step Scale Estimator) [23]; they employ the mean shift technique to model the inlier distribution and obtain a corresponding scale, and also they enable RANSAC to be data driven, but at the cost of make the whole process more time consuming than the original proposal. The mentioned modifications to original RANSAC focused mainly in improvements to the objective function; nevertheless, all of them consider that the estimation process is approached as an optimization problem where the search strategy is a random walking algorithm while the objective function is fixed to the number of inliers associated to the candidate model. In order to overcome this kind of search, in this chapter is used an intelligent search, considering that the problem at hand is an optimization problem, which can be solved by an intelligent search in the solution space. Two important difficulties in selecting a search strategy for RANSAC are the high multi-modality and the complex characteristics of the estimation process produced by the elevated contamination of the dataset. Under such circumstances, classical methods present a bad performance [24, 25], making way for recent new approaches that have been proposed to solve complex and ill-posed engineering problems. These methods include the application of modern optimization techniques such as evolutionary algorithms and metaheuristic techniques [26, 27] which have delivered better solutions over those obtained by classical methods.

A well know evolutionary algorithm is the so called Harmony Search (HS) [28]. This algorithm is inspired in the improvisation process that an orchestra achieves when searches for harmony, which is constantly improved. Just as the metaphor, this algorithm produces new solutions based on the previous candidates. In the algorithm, the candidate solutions are vectors that represents harmonies, and whose improvements in the evolutionary process are the improvisations of a real orchestra. Compared with other metaheuristics, HS has fewer mathematical requisites [29, 30]. Moreover, this algorithm has demonstrated a better convergence in the search for the optimum than other algorithms, such as Genetic Algorithms [31, 32]. For those reasons, this algorithm has been applied to solve a wide range of practical optimization problems such as structural optimization [33], parameter estimation of the nonlinear Muskingum model [34], design optimization of water distribution networks [35], vehicle routing [36], image segmentation [37], circle detection in images [38], among others. An improvement in HS is the local search parameter (BW) dynamically adjustment, proposed to improve the balance between exploration and exploitation during the search process (see [29]). This is because the original algorithm underperforms in local searching [39–42], particularly in parameter identification applications. Nevertheless, as that adjustment follows an exponential function, then longer exploitation times are employed, and therefore the exploring capacity of the algorithm is damaged. Accordingly, in [43] it was proposed the use of a linear model for the dynamic adjustment, which has better searching capacities than the approaches based on exponential functions. Such approach, in combination with the original RANSAC method, is utilized in this chapter for the robust estimation of multiple view relations from point correspondences. In that sense, the method considers a different sampling strategy to generate putative solutions, under which new candidate solutions are generated by considering previous candidate solutions and their corresponding quality, rather than in a purely random manner. The main advantage in using that approach is a considerable reduction in the number of iterations, without a sacrifice of the robust capabilities of RANSAC. The use of HSRANSAC is compared against other methods to generate view relations from point correspondences; moreover, it is applied to solve the problem of position estimation of a humanoid robot, obtaining good results in that engineering application.

The remaining of the chapter is as follows: Sect. 9.2 gives details about the problem of image matching when multiple views are used. Section 9.3 introduces the RANSAC method, whereas Sect. 9.4 show describes the HS metaheuristic approach; in the same venue, Sect. 9.5 summarizes the utilized algorithm: HSRANSAC. The experimental part is explained in Sect. 9.6, exposes an application of the method in robotics. Finally, Sect. 9.7 gives some remarks, and draws some final conclusions.

9.2 Geometric Transformation

A known problem in image processing is the image matching, which consists in finding a geometric transformation capable of mapping an image taken from a given scene, to another image taken from the same scene, but from another point of view. In fact, to determine the correspondence among points, it is necessary to find corresponding points on both images; to that end, it is necessary the utilization of automatic detection and matching algorithms [44, 45]. Usually the utilized points are arranged in vectors of parameters, or descriptors, which do not always allow a complete discrimination, giving as a result an erroneous matching about the correspondence of points located on different parts of different images.

This section deals with the geometric relations of points between two viewsem, considering the case of homography.

Assume that there is a collection of pairs of the corresponding points that are found on two images

$$\mathbf{U} = \left\{ \left(\mathbf{x}_1, \mathbf{x}_1'\right), \left(\mathbf{x}_2, \mathbf{x}_2'\right), \ldots, \left(\mathbf{x}_M, \mathbf{x}_M'\right) \right\}, \tag{9.1}$$

where $\mathbf{x}_i = (x_i, y_i, 1)^T$ and $\mathbf{x}_i' = (x_i', y_i', 1)^T$ are the positions of points in the first and second image, respectively.

Two images of the same scene but with different perspective can be geometrically linked through a plane Q of the scene by a homography matrix $\mathbf{H} \in \mathbb{R}^{3 \times 3}$ (Fig. 9.1). Such transformation connects corresponding points of the plane projected into two images either by $\mathbf{x}_i' = \mathbf{H}\mathbf{x}_i$ or by $\mathbf{x}_i = \mathbf{H}^{-1}\mathbf{x}_i'$. In order to calculate the homography that links the two views, it is necessary to solve a linear system obtained from a set of four point matches [46]; such homography is later evaluated,

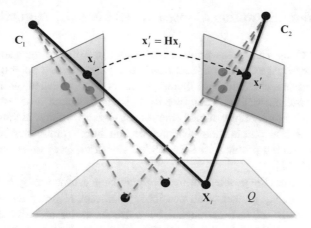

Fig. 9.1 Homography between two views

to measure its quality, by considering the distance between the position of the point calculated and the actually observed position. Therefore, the mismatch error EH_i^2 produced by the i-correspondence $(\mathbf{x}_i, \mathbf{x}_i')$ is defined as the sum of squared distances from the points to their estimated positions:

$$EH_i^2 = \left[d(\mathbf{x}_i', \mathbf{H}\mathbf{x}_i) \right]^2 + \left[d(\mathbf{x}_i, \mathbf{H}^{-1}\mathbf{x}_i') \right]^2, \qquad (9.2)$$

where $\eta = d(\mathbf{x}_i, \mathbf{H}^{-1}\mathbf{x}_i')$ and $\eta' = d(\mathbf{x}'_i, \mathbf{H}\mathbf{x}_i)$ correspond to the errors produced in the first and second image, respectively.

An example of an error evaluation is depicted in Fig. 9.2; in this case, five correspondence points $\mathbf{U} = \left\{ (\mathbf{x}_1, \mathbf{x}_1'), \ldots, (\mathbf{x}_5, \mathbf{x}_5') \right\}$ are distributed in both views. Accordingly, the correspondence presents a big error, since the distances (η, η') between the points $(\mathbf{x}_3, \mathbf{x}_3')$ are considerably big.

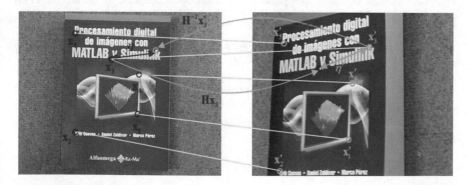

Fig. 9.2 An example of a homography's evaluation

9.3   Random Sampling Consensus (RANSAC) Algorithm

The RANSAC algorithm was proposed as a tool to estimate the geometric transformation from image correspondences over two views, with a potential significant
number of mismatches among them, while correct matches will follow the
homography transformation. Therefore, the aim is to obtain a set of *inliers* consistent with the homography transformation by using a robust technique, leaving
outside that set the points inconsistent with the homography transformation. In
order to solve such a problem, the RANSAC algorithm has proven to be the most
successful [15–17].

RANSAC solves the problem of model parameters estimation by finding the best
hypothesis h^B among the set of all possible hypotheses H generated by the source
data, which is typically contaminated by noise. In order to build the candidate
hypothesis h_i about the unknown parameters, a sample $\mathbf{S}_i$ of a minimum size (s) is
required for model estimation (for example, a sample of only two points is sufficient
to calculate a straight line, $s = 2$ and four to obtain a homography $s = 4$). Under this
consideration, the probability of finding an outlier is reduced. Considering that the
number of elements contained in a sample is small, the amount of possible samples
that can be generated from the complete source data $\mathbf{U}$ is enormous. Under such
circumstances, the exhausting testing of all samples for a reasonable time is
impossible. RANSAC face such problem because it only considers G samples
which are randomly selected and evaluated. Algorithms of the RANSAC family
consist of G iterations of the following cycle:

(1)	Construct a sample $\mathbf{S}_i \subset \mathbf{U}$ consisting of s different elements
(2)	Build the hypothesis h_i based on the sample $\mathbf{S}_i$
(3)	Evaluate the degree of agreement A_i of the hypothesis h_i with the set of all source data $\mathbf{U}$

Once all the G hypotheses have been constructed and evaluated, then the
hypothesis h^B with the best degree of agreement is chosen, which can be considered
a robust estimate of the model parameters. That operation can be described as an
optimization problem as follows:

$$h^B = \arg\max_{i=1,\ldots,G} A_i(\mathbf{U}, h_i) \tag{9.3}$$

In the same order of ideas, the maximization of the degree of agreement (number
of inliers) is equivalent to the minimization of the penalty function whose value
depends on the number of outliers. The degree of agreement $A_i(\mathbf{U}, h_i)$ is calculated
as follows:

$$A_i(\mathbf{U}, h_i) = \sum_{j=1}^{M} \theta\left(e_j^2(h_i)\right), j = 1, \ldots, M, \quad \theta\left(e_j^2(h_i)\right) = \begin{cases} 0 & e_j^2(h_i) > Th \\ 1 & e_j^2(h_i) \le Th \end{cases},$$

$$(9.4)$$

considering that Th is the permissible error, M the number of elements contained in the source data U, $e_j^2(h_i)$ is the quadratic error produced by the jth data considering the hypothesis h_i, and $e_i^2(h_i)$ corresponds to EH_i^2 which represents the error produced by the ith correspondence.

The hypothesis with the maximum degree of agreement is considered as the best matching criterion; in the original RANSAC algorithm, the number of inliers determines the quality of the hypotheses: for a given value of the permissible error Th, the point j that produces the error $e_j^2(h_i)$ is regarded to be an inlier of h_i if its value does not exceed the threshold Th, otherwise the point is regarded as an outlier. Moreover, the optimal hypothesis h^B is found and the penalty is minimized by means of a search strategy based on random walking; therefore many attempts are necessary to investigate in sufficient detail the space of possible samples and to find the sample for which the hypothesis has the greatest degree of agreement on the source data. The number of iterations and thus the time spent for the search can be reduced by choosing points according to some directed rules, rather than randomly. Optimization algorithms have been considered as a robust scheme in contrast to the random search [46]. In an optimization algorithm, new candidate solutions are generated in accordance to information obtained from past candidate solutions.

Accordingly, in this chapter it is used the HS algorithm together with the original RANSAC algorithm, with the idea of testing minimum-sized samples with the directed search inspired by the improvisation process that occurs when a musician searches for a better state of harmony; by using that approach, it is performed an efficient search among the correspondences to generate models of higher quality.

9.4 Harmony Search Algorithm

The candidate solution is represented as an n-dimensional real vector and it is called a 'harmony' in the HS algorithm. In this algorithm the population is considered the Harmony Memory (HM), and at the beginning is randomly generated. In order to generate a new candidate solution it is taken an element of the HM, and either a random resetting- or a pitch adjustment- operation is applied. After such operation, the HM is updated by considering the new generated candidate harmony, and the worst element in the HM: the last is replaced in case that the new harmony is better than such element. Such process is iterated until a criterion is reached. In its basic form, the algorithm has three phases: initialization, improvisation, and updating. Next section explains each one of them.

9.4.1   Initialization, Improvisation, and Updating

A global optimization problem can be stated as follows: min $f(\mathbf{p})$: $p(j) \in$ $[l(j), u(j)], j = 1, 2, \ldots, n$, where $f(\mathbf{p})$ is the objective function, $\mathbf{p} = (p(1), p(2), \ldots, p(n))$ is the set of design variables, n is the number of design variables, and $l(j)$ and $u(j)$ are the lower and upper bounds for the design variable $p(j)$, respectively. In the algorithm, besides the definition of the optimization problem, it is also necessary to define some other parameters, such as: the memory size (HMS), the harmony-memory consideration rate (HMCR), the pitch adjusting rate (PAR), the distance bandwidth (BW), and the number of improvisations (NI), which is the number of algorithm iterations. It is important to point out that the correct adjustment of those parameters is fundamental to an adequate convergence of the algorithm.

The first step in the algorithm is the memory initialization; in this stage, initial vector components at HM, i.e. *HMS* vectors, are configured. Let $\mathbf{p}_i = \{p_i(1), p_i(2), \ldots, p_i(n)\}$ represent the ith randomly-generated harmony vector: $p_i(j) = l(j) + (u(j) - l(j)) \cdot \text{rand}(0, 1)$ for $j = 1, 2, \ldots, n$ and $i = 1, 2, \ldots, HMS$, where rand(0, 1) is a uniform random number between 0 and 1. Then, the HM matrix is filled with the *HMS* harmony vectors as follows:

$$\text{HM} = \begin{bmatrix} \mathbf{p}_1 \\ \mathbf{p}_2 \\ \vdots \\ \mathbf{p}_{HMS} \end{bmatrix} \tag{9.5}$$

The next step is the improvisation of new candidate harmonies. In this phase, a new harmony vector $\mathbf{p}_{new}$ is built by applying the following three operators: memory consideration, random re-initialization and pitch adjustment. Generating a new harmony is known as 'improvisation'. In the memory consideration step, the value of the first decision variable $p_{new}(1)$ for the new vector is chosen randomly from any of the values already existing in the current HM i.e. from the set $\{p_1(1), p_2(1), \ldots, p_{HMS}(1)\}$. For this operation, a uniform random number r_1 is generated within the range [0, 1]. If r_1 is less than *HMCR*, the decision variable $p_{new}(1)$ is generated through memory considerations; otherwise, $p_{new}(1)$ is obtained from a random re-initialization between the search bounds $[l(1), u(1)]$. Values of the other decision variables $p_{new}(2), p_{new}(3), \ldots, p_{new}(n)$ are also chosen accordingly. Therefore, both operations, memory consideration and random re-initialization, can be modelled as follows:

$$p_{new}(j) = \begin{cases} p_i(j) \in \{p_1(j), p_2(j), \ldots, p_{HMS}(j)\} & \text{with probability } HMCR \\ l(j) + (u(j) - l(j)) \cdot \text{rand}(0, 1) & \text{with probability } 1 - HMCR \end{cases}$$
$$\tag{9.6}$$

Every component obtained by memory consideration is further examined to determine whether it should be pitch-adjusted. For this operation, the Pitch-Adjusting

Rate (*PAR*) is defined as to assign the frequency of the adjustment and the Bandwidth factor (*BW*) to control the local search around the selected elements of the HM. Hence, the pitch adjusting decision is calculated as follows:

$$p_{new}(j) = \begin{cases} p_{new}(j) = p_{new}(j) \pm \text{rand}(0,1) \cdot BW & \text{with probability } PAR \\ p_{new}(j) & \text{with probability } (1 - PAR) \end{cases}$$

$$(9.7)$$

Pitch adjusting is responsible for generating new potential harmonies by slightly modifying original variable positions. Such operation can be considered similar to the mutation process in evolutionary algorithms. Therefore, the decision variable is either perturbed by a random number between $-BW$ and BW or left unaltered. In order to protect the pitch adjusting operation, it is important to assure that points lying outside the feasible range $[l, u]$ must be re-assigned i.e. truncated to the maximum or minimum value of the interval.

Finally, the last step in the algorithm is the updating of the harmony memory: once a new harmony vector $\mathbf{p}_{new}$ is generated, the harmony memory is updated considering a competition between this individual, and the worst of the population, $\mathbf{p}_w$. Therefore $\mathbf{p}_{new}$ will replace $\mathbf{p}_w$ and become a new member of the HM in case the fitness value of $\mathbf{p}_{new}$ is better than the fitness value of $\mathbf{p}_w$.

The computational procedure is shown in the next pseudocode:

Step 1:	Set the parameters *HMS*, *HMCR*, *PAR*, *BW* and *NI*
Step 2:	Initialize the HM and calculate the objective function value of each harmony vector

Step 3:	Improvise a new harmony $\mathbf{p}_{new}$ as follows:
	for ($j = 1$ to n) do
	    if ($r_1 <$ *HMCR*) then
	        Select randomly a number a where $a \in (1, 2, \ldots, HMS)$
	        $p_{new}(j) = p_a(j)$
	        if ($r_2 <$ *PAR*) then
	            $p_{new}(j) = p_{new}(j) \pm r_3 \cdot BW$ where $r_1, r_2, r_3 \in \text{rand}(0,1)$
	        end if
	        if $p_{new}(j) < l(j)$
	            $p_{new}(j) = l(j)$
	        end if
	        if $p_{new}(j) > u(j)$
	            $p_{new}(j) = u(j)$
	        end if
	    else
	        $p_{new}(j) = l(j) + r \cdot (u(j) - l(j))$, where $r \in \text{rand}(0,1)$
	    end if
	end for

Step 4:	Update the HM as $\mathbf{p}_w = \mathbf{p}_{new}$ if $f(\mathbf{p}_{new}) > f(\mathbf{p}_w)$
Step 5:	If *NI* is completed, the best harmony vector $\mathbf{p}_b$ according to its fitness value in the HM is returned; otherwise go back to step 3

9.4.2 Dynamic Linear Adjustment of Bandwidth

As in other metaheuristic algorithms, in HS the balance between exploration and exploitation is a desirable feature. Exploration is the process of visiting entirely new points of a search space, whereas exploitation is the process of refining those points within the neighborhood of previously visited locations in order to improve their solution quality. In the case of the standard HS, such equilibrium is achieved with the parameter BW: a large value enables the algorithm to achieve exploration of the search space, while a small value permits a fine-grain search in the search space. In this chapter it is used a BW dynamically adjusted, that favors exploration at early stages, and enables the exploitation in the last stages of the algorithm. The adjustment uses a linear model defined as follows:

$$
BW(k) = \begin{cases} BW_{\max} - \left(\frac{BW_{\max}-BW_{\min}}{2 \cdot NI}\right) \cdot 3k & \text{if } k < \left(\frac{2}{3}\right) NI \\ BW_{\min} & \text{if } k \geq \left(\frac{2}{3}\right) NI \end{cases} \tag{9.8}
$$

where k is the iteration index, while and are the maximum and minimum BW values, respectively. In contrast to exponential adjustment [26], linear models, as the one used in this chapter, allow a better balance between exploration and exploitation (fine-tuning) of the search process [40].

9.5 Geometric Estimation with Harmony Search

The candidate solution correspond to a sample of length s from a set consisting of M elements. In the original RANSAC algorithm the search strategy used is a random walk; therefore, the use of HS is based on the quality of the found samples, rather than a pure random search. In the methodology utilized in this chapter, the quality of every sample is obtained by the matching degree $f(\mathbf{p}_i)$ with the hypothesis h_i that is constructed based on the correspondence numbers coded within $\mathbf{p}_i$. In the same order of ideas, it is necessary to obtain the parameters of $\mathbf{H}$ through a set $\mathbf{U} = \{(\mathbf{x}_1, \mathbf{x}'_1), (\mathbf{x}_2, \mathbf{x}'_2), \ldots, (\mathbf{x}_M, \mathbf{x}'_M)\}$ of M different correspondences; this can be achieved with the next algorithm:

1. Configuration		
	(a)	Set the parameters HMS, $HMCR$, PAR, $BW_{\min}, BW_{\max}$ and NI
2. Initial population		
	(a)	Build the harmony memory (HM) HM $= \{\mathbf{p}_1, \mathbf{p}_2, \ldots, \mathbf{p}_{HMS}\}$ where each individual $\mathbf{p}_i$ consists 4 random non-repeating indices from 1 to M
	(b)	Compute homography $\mathbf{H}_i$ (hypothesis h_i) by using the indices from $\mathbf{p}_i$

(continued)

(continued)

(c)	Calculate the fitness value $f(\mathbf{p}_i)$ as the matching quality of the constructed homography $\mathbf{H}_i$ considering the whole available data $\mathbf{U}$. Such fitness value is calculated by using a new objective function defined as: $$F(\mathbf{a}_i(k)) = \sum_{j=1}^{M} \theta\left(e_j^2(h_i)\right) - \lambda \cdot e_j^2(h_i)$$ $$\theta\left(e_j^2(h_i)\right) = \begin{cases} 0 & e_j^2(h_i) > Th \\ 1 & e_j^2(h_i) \le Th \end{cases} \qquad (9.9)$$ where $e_j^2(h_i)$ represents the quadratic error produced by the jth correspondence considering the hypothesis h_i whereas λ is the penalty associated with the mismatch magnitude. Such error corresponds to the mismatch EH_j^2 generated by the evaluation of $\mathbf{H}_i$

3. Iterations $k = 1, \ldots, NI$

(a)	Generate a new harmony $\mathbf{p}_{new}$ (candidate solution) as follows:
	
	$$BW(k) = \begin{cases} BW_{max} - \left(\frac{BW_{max} - BW_{min}}{2 \cdot NI}\right) \cdot 3k & \text{if } k < \left(\frac{2}{3}\right)NI \\ BW_{min} & \text{if } k \ge \left(\frac{2}{3}\right)NI \end{cases}$$
	
	for ($j = 1$ to n) do
	  if ($r_1 < HMCR$) then
	    Select randomly a number a where $a \in (1, 2, \ldots, HMS)$
	    $p_{new}(j) = p_a(j)$
	    if ($r_2 < PAR$) then
	      $p_{new}(j) = p_{new}(j) \pm r_3 \cdot BW(k)$ where $r_1, r_2, r_3 \in \text{rand}(0, 1)$
	    end if
	    if $p_{new}(j) < l(j)$
	    $p_{new}(j) = l(j)$
	    end if
	    if $p_{new}(j) > u(j)$
	    $p_{new}(j) = u(j)$
	    end if
	  else
	    $p_{new}(j) = l(j) + r \cdot (u(j) - l(j))$, where $r \in \text{rand}(0, 1)$
	  end if
	end for

(b)	Compute homography $\mathbf{H}_i$ by using the indices from $\mathbf{p}_{new}$
(d)	Calculate the fitness value $f(\mathbf{p}_{new})$ as the matching quality of the constructed homography $\mathbf{H}_{new}$ considering the whole available data $\mathbf{U}$. Such fitness value is calculated by using the objective function described in Eq. 9.4
(e)	Update the HM as $\mathbf{p}_w = \mathbf{p}_{new}$ if $f(\mathbf{p}_{new}) > f(\mathbf{p}_w)$

4. Estimation result

(a)	The best estimation $\mathbf{H}^B$ consists of the parameters computed by using the indices from the best element $\mathbf{p}^B$ of HM in terms of its affinity, so that $\mathbf{p}^B = \arg\max_{i=1,\ldots,HMS} f(\mathbf{p}_i)$

As can be seen, the approach considers a combination of RANSAC with HS, and therefore, it considers the RANSAC as an optimization problem; accordingly, even though other objective functions could be incorporated for a more accurate evaluation of the quality of the candidate model, in this work it is used the expression given in Eq. 20. Such function, in contrast to the one considered in traditional RANSAC, includes not only the number of inliers, but also the approximation error; in such sense, the found solution could be considered a best trade-off between both objectives, which are usually in disagreement. Therefore, this methodology can significantly reduce the number of iterations while are preserved the capabilities of the RANSAC method.

9.6 Experimental Results

In the experimental part the results were obtained by considering, on one hand, the effect of the HS parameters in the estimation results, and on the other hand, the results obtained with the method when applied over synthetic data and over real images. Also, the performance indexes considered were: the number of inliers ($NofI$), the error (E_s, E_r) and the number of function evaluations (NFE), which are related with the solution accuracy, and with the computational cost, respectively. The $NofI$ expresses the number of elements contained in the set I of detected inliers, whereas E_s, E_r provides a quality measure of the estimated relation; for the synthetic data, such value is determined with:

$$E_s = \left(\sum_{ij} \frac{d^2(\mathbf{x}_i^j, \hat{\mathbf{x}}_i^j)}{NofI} \right)^{1/2}, \, i \in \mathbf{I}, \, j \in \{1, 2\}, \qquad (9.10)$$

where $\mathbf{x}_i^j$ is the inlier point calculated by the estimated relation in the j-view, $\hat{\mathbf{x}}_i^j$ is the inlier ground true point and $d(\cdot)$ is the Euclidian distance between the points. Therefore, E_s evaluates the fit of the estimated relation, computed from the noisy data, against the known ground truth points. On the other hand, for real data the error is calculated by considering the standard deviation of the inliers:

$$E_r = \left(\sum_i \frac{e_i^2}{NofI} \right)^{1/2}, \, i \in \mathbf{I}, \qquad (9.11)$$

considering that e_i^2 is the quadratic error produced by the ith inlier and, in this chapter, it is also equivalent to EH_i^2. Finally, with regards to NFE, this value specifies the total number of transformations that have been evaluated by the algorithm until the best estimation has been reached.

9.6.1 Effects of HS

In this part, two parameters were considered as the most important [47], and therefore tested, for a good performance of the HS algorithm: the harmony-memory

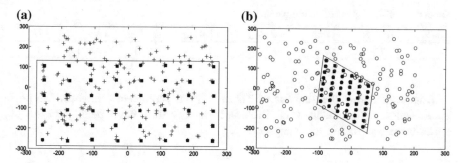

Fig. 9.3 Experimental setup for HS parameters test, where the black squares represent the detected inliers

consideration rate (*HMCR*) and pitch adjusting rate (*PAR*). In order to accomplish such testing, several values of those parameters were probed to the calculus of a synthetic homography; accordingly, it was used a rectangular pattern of 8 × 6 elements within a 2-dimensional space of [−300, 300], which was later transformed by a random homography. Moreover, normally distributed noise was added to the correspondences in the second view. Conducive to illustrate the experiments, Fig. 9.3 show the first and second view, respectively; in this case, the black squares indicate the position in the first view of a point from the second view as a result of the **H** transformation. Likewise, the black squares in Fig. 9.3b exhibit the position in the second view of a point from the first view as a result of the **H** transformation.

In the experiment, the maximum number of iterations is set to 950. *HMS*, BW_{max}, BW_{min}, λ and *Th* are fixed to 50, 10, 1, 0.001 and 5, respectively. The results report the number of inliers (*NofI*) and the produced estimation error (E_s) of HS-RANSAC, averaged over 30 runs, for the different values of *HMCR* and *PAR*. In the experiment, the parameter values are modified considering specific interval. *HMCR* varies from 0.5 to 0.8 whereas *PAR* changes from 0.1 to 0.4. The outcomes, shown in Table 9.1, suggest that a proper combination of different parameter values can improve the performance of HS-RANSAC, and therefore the quality of the estimations. The best parameter configuration in the experiment is highlighted in Table 9.1.

Table 9.1 Effect of the HS parameters in the estimation process

	(*NofI*, E_s)			
	HMCR = 0.5	*HMCR* = 0.6	*HMCR* = 0.7	*HMCR* = 0.8
PAR = 0.1	(21, 4.2147)	(26, 3.8124)	(29, 3.4721)	(27, 4.0112)
PAR = 0.2	(22, 4.1457)	(35, 2.0974)	(36, 2.1474)	(31, 3.3784)
PAR = 0.3	(21, 4.5714)	(38, 1.1124)	**(40, 0.8514)**	(35, 2.0053)
PAR = 0.3	(23, 4.0781)	(34, 2.0078)	(37, 2.0012)	(31, 3.4079)

Table 9.2 HS-RANSAC estimator parameters

HMS	HMCR	PAR	BW_{max}	BW_{min}	NI	λ	Th
50	0.7	0.3	10	1	950	0.001	5

After considering the analysis of Table 9.1, the parameter values for the proposed estimator are defined in Table 9.2; such values have been kept in all experiments reported in this chapter.

9.6.1.1   Results Over Synthetic and Real Homographies

The HS-RANSAC was compared against other well known methods for homography estimation over synthetic as well as over real data: RANSAC [14], the MLESAC [20], the SIMFIT method [21], the projection-pursuit algorithm [22], the TSSE [23] and the PSO algorithm (PSO-RANSAC) [48]. The first five are RANSAC-based estimators, and their results are broadly known. Also, those algorithms were tuned based on the references. In the case of PSO-based, this was included only as a reference, to validate the performance of the HS as an optimization approach. In that sense, and in order to conduct a fair comparison between the HS version used in this work and PSO, it has been also chosen an enhanced version of PSO with similar characteristics. Therefore, it is used, in the comparisons, the PSO version reported in [48]. Such approach is proposed to mitigate the premature convergence problem of the original PSO method. It incorporates two new elements: (1) a weight factor w and (2) a constriction factor V_{max}. Similar to BW in the HS method, the weight factor w is linearly decreased during the algorithm execution to regulate the attraction force towards the best particle seen so-far. On the other hand, the constriction factor V_{max} permits to limit the particle velocities in order to control their trajectories. Under such circumstances, the enhanced PSO version is used in combination with RANSAC considering the following configuration: $P = 10$, $c_1 = 2$, $c_2 = 2$, $Th = 5$ whereas the weight factor w decreases linearly from 0.9 to 0.2. Additionally, the constriction factor V_{max} is fixed to 2; such configuration presents the best possible performance [48].

9.6.1.2   Homography Estimation with Synthetic Data

In this part, the reported results correspond to the homography estimation when it is considered synthetic data (see Sect. 9.6.1), although in this case the incorporated outliers varies from 0 to 100%.

The performance of each algorithm was averaged over 50 runs for each one (Fig. 9.4). In the comparison analysis it was used a concept called breakdown point [18]. The breakdown point is identified as the highest outlier ratio from which the algorithm degrades its capacity to find inliers. It can be seen from Fig. 9.4a that

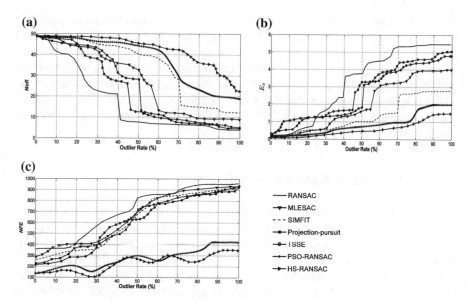

Fig. 9.4 Results for the estimation of **H**, over synthetic data

standard RANSAC has a breakdown point at 40%, the MLESAC at 55%, the SIMFIT method at 70%, the projection-pursuit algorithm at 50%, the TSSE at 45% and the PSO-RANSAC at 80%. In contrast to such methods, the proposed approach, HS-RANSAC, does not seem to have a prominent breakdown point, since its capacity to detect inliers smoothly degrades. It is also observed that the HS-RANSAC algorithm presents the best performance in terms of the number of inliers (*NofI*), as it is able to detect most of them. For the estimated **H**, the error E_s (Fig. 9.4b) is fairly comparable for all methods until they reach their breakdown points. Nonetheless, the proposed algorithm performed better, being the only algorithm that consistently found the minimum error at all outlier ratios.

As can be seen from Fig. 9.4c, in terms of functions evaluations (NFE), the algorithms RANSAC, MLESAC, the projection pursuit, and the TSSE, almost invest the same number of iterations to find the best estimation of H; this behavior can be explained by the random walking used by those techniques as a search procedure. On the other hand, the PSO-RANSAC and the HS-RANSAC (which consider the search strategy as an optimization problem) maintain a considerably low *NFE* value with independence from the number of outliers. In such sense, it is clear from these experiments that the use of an optimization approach can considerably reduce the *NFE* value. Nevertheless, if considered the multimodality and complex characteristics of the estimation process over a contaminated dataset, then not every algorithm is suitable to solve the estimation problem. Therefore, although the PSO-RANSAC finds its best estimated fundamental matrix **H** investing approximately the same number of evaluations as the HS-RANSAC, such estimated matrix represents only a sub-optimal solution. This fact can be observed in

Fig. 9.4b where it is clear that the PSO-RANSAC algorithm presents higher E_s values in comparison to the HS-RANSAC approach. The reason of this problem points to those operators used by PSO for modifying the individual positions. In PSO, during their evolution, the position of each agent in the next iteration is updated yielding an attraction towards the position of the best particle seen so-far.

Such behavior produces that the entire population, as the algorithm evolves, concentrates around the best particle, favoring the premature convergence (reaching sub-optimal solutions) [49].

9.6.1.3 Homography Estimation with Real Images

In this section, the experimental results of the estimation of homographies **H** considering real images is reported. To evaluate the estimation performance of the proposed method, Table 9.3 tabulates the comparative inlier detection performance of the standard RANSAC [14], the MLESAC [20], the SIMFIT method [21], the projection-pursuit algorithm [22], the TSSE [23], the PSO algorithm (PSO-RANSAC) [48] and the HS-RANSAC approach, in terms of the detection rate (DR), the error (E_r) and the number of function evaluations (NFE). The experimental data set includes 4 images (Images A, B, C and D) which are shown in Figs. 9.5, 9.6, 9.7 and 9.8. Such images contain a determined number of inliers which have been detected and counted by a human expert (A = 86, B = 72, C = 56 and D = 122). Such values act as ground truth for all the experiments. For the comparison, the detection rate (DR) is defined as the ratio between the number of inliers correctly detected by the algorithm (NofI value) and the total number of inliers determined by the expert. The results consider 50 different executions for each algorithm over the four images. Experimental results show that the proposed HS method accomplishes at least a 94.2% of inlier detection accuracy. A close inspection of Table 9.3 also reveals that the proposed approach is able to achieve the smallest error (E_r), yet requiring a few number of function evaluations (NFE) for most cases. Figures 9.5, 9.6, 9.7 and 9.8 also exhibit the results after applying the HS-RANSAC estimator. Such results present the median case obtained throughout 50 runs.

9.6.2 Estimation of Position in a Humanoid Robot

In this section it is presented an interesting application of HSRANSAC to estimate the position of an humanoid robot.

In the last decades, much work has already been accomplished in the area of humanoid robotics [50, 51]. Position determination for humanoid robots is a critical problem, since it is used to control their balance and locomotion. Recently, a notable research [52] has been devoted to achieve better performance in system position for humanoid robots by using sensor fusion methods. In general,

Table 9.3 Inlier detection comparison considering four test images

Image	Method	*NofI*	Missing	False alarms	DR (%)	E_r	*NFE*
(A) Total number of Inliers **86**	Standard RANSAC	41	45	21	47.7	4.75	876
	MLESAC	55	31	14	63.9	3.11	852
	SIMFIT	62	24	11	72.0	2.98	842
	Projection-pursuit	58	28	12	67.4	3.53	798
	TSSE	48	38	14	55.8	3.42	815
	PSO-RANSAC	75	11	8	87.2	1.68	491
	HS-RANSAC	82	4	5	95.3	0.88	396
(B) Total number of Inliers **72**	Standard RANSAC	32	40	18	44.4	3.98	765
	MLESAC	40	32	14	55.5	3.43	825
	SIMFIT	58	14	8	80.5	2.87	891
	Projection-pursuit	47	25	12	65.2	3.12	759
	TSSE	43	29	16	59.7	3.47	786
	PSO-RANSAC	63	9	5	87.5	1.51	374
	HS-RANSAC	70	2	3	97.2	0.79	328
(C) Total number of Inliers **56**	Standard RANSAC	24	32	15	42.8	2.96	689
	MLESAC	27	29	11	48.2	2.41	628
	SIMFIT	42	14	9	75.0	1.98	724
	Projection-pursuit	37	19	13	66.0	2.85	754
	TSSE	32	24	14	57.1	2.74	776
	PSO-RANSAC	48	8	9	85.7	0.94	349
	HS-RANSAC	53	3	5	94.6	0.25	272
(D) Total number of Inliers **122**	Standard RANSAC	62	60	22	50.8	4.02	832
	MLESAC	77	45	18	63.1	3.41	924
	SIMFIT	90	32	13	73.7	2.86	845
	Projection-pursuit	75	47	19	61.4	3.52	914
	TSSE	76	46	21	62.2	3.73	887
	PSO-RANSAC	110	12	10	90.1	1.41	427
	HS-RANSAC	115	7	5	94.2	0.51	338

integrating information from different sensors increases the versatility of the system, but also its cost and complexity. Vision is one of the most studied sensory modalities for position and navigation purposes since it provides rich information of the environment.

The framework of the approach presented in this section, as an application, is a vision system consisting of a fixed camera mounted on a Bioloid© humanoid robot. In the approach, the position (x, y) of the robot is computed considering the homography estimated by the HS-RANSAC. Therefore, the idea is to calculate the

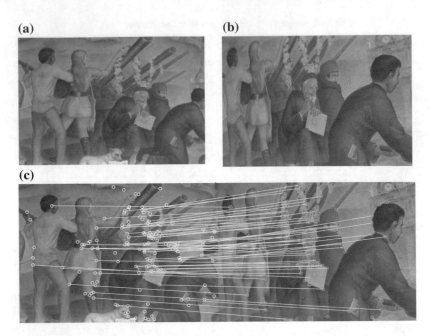

Fig. 9.5 Test image "A": **a** first view, **b** second view, **c** correspondence points and inliers produced by HS-RANSAC

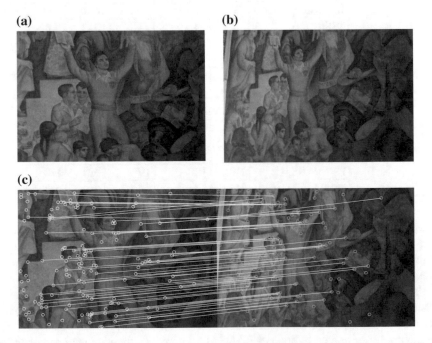

Fig. 9.6 Test image "B": **a** first view, **b** second view, **c** correspondence points and inliers produced by HS-RANSAC

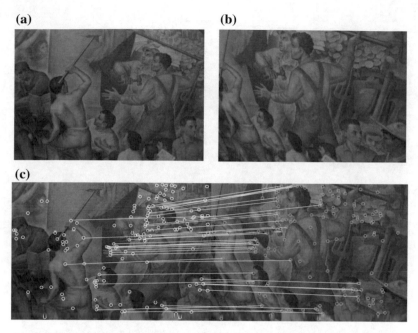

Fig. 9.7 Test image "C": **a** first view, **b** second view, **c** correspondence points and inliers produced by HS-RANSAC

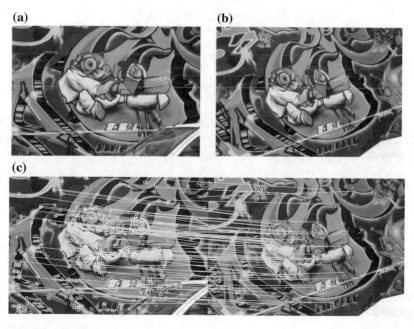

Fig. 9.8 Test image "D": **a** first view, **b** second view, **c** correspondence points and inliers produced by HS-RANSAC

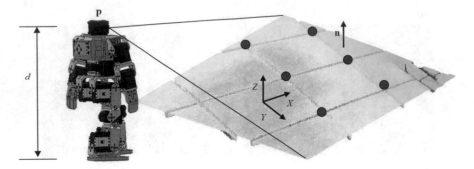

Fig. 9.9 Process of planar motion calculation based on homographies

planar motion of the humanoid robot through the estimated homographies. Figure 9.9 illustrates the process of planar motion calculation.

The homography can be related to camera motion and plane location as follows:

$$\mathbf{H} = \mathbf{R} + \frac{1}{d}\mathbf{t}^T\mathbf{n}, \tag{9.12}$$

where d is the distance from the camera to the plane (the height of the humanoid approximately). $\mathbf{R}$ describes a rotation γ about the Z axis and can be expressed as:

$$\mathbf{R} = \begin{bmatrix} \cos\gamma & \sin\gamma & 0 \\ -\sin\gamma & \cos\gamma & 0 \\ 0 & 0 & 1 \end{bmatrix}, \tag{9.13}$$

And $\mathbf{t}$ is a translation vector with the form:

$$\mathbf{t} = (t_x, t_y, 0), \tag{9.14}$$

As the unit normal $\mathbf{n}$ is $(0, 0, 1)$, considering the point $\mathbf{p}$, the rotation matrix $\mathbf{R}$ and the vector $\mathbf{t}$ (where $\mathbf{R}$ and $\mathbf{t}$ are calculated from the homography $\mathbf{H}$), the new planar position $\mathbf{p}_{new}$ can be computed as:

$$\mathbf{p}_{new} = \mathbf{R}\mathbf{p} + \mathbf{t} \tag{9.15}$$

More details about planar motion on homography can be found in the specialized literature. The HS-RANSAC algorithm and Eqs. 9.12–9.15 were implemented in a Raspberry Pi. Since the computation must be verified in real time, the number of iterations is fixed to only 150. Figure 9.10 shows the calculated positions from the homographies estimated during the humanoid locomotion. Such a Figure demonstrates that the information of the estimated position adequately reflexes the humanoid movement in spite of the reduced number of iterations.

Fig. 9.10 Position calculated
from the homography

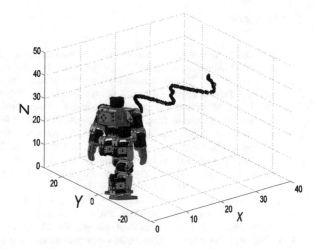

9.7 Conclusions

In this chapter were compared six methods for estimating homographies from point correspondences, being two of them evolutionary-based: PSORANSAC, and HSRANSAC. Such approaches combines the original RANSAC method with the search capabilities of evolutionary algorithms; in the first case, the method was combined with Particle Swarm Optimization, whereas the second one was combined with Harmony Search. Under the new mechanism, new candidate solutions are generated iteratively by taking into consideration the quality of models produced by previous candidate solutions, instead of relying over a pure random selection as it is the case of RANSAC. On the other hand, a more accurate objective function was incorporated to adequately asses the quality of a candidate model. Such approaches can substantially reduce the number of iterations still preserving the robust capabilities of RANSAC. The other compared techniques were the standard RANSAC, the MLESAC, the SIMFIT method, the projection-pursuit algorithm, and the TSSE. The efficiency of each algorithm was evaluated in terms of the detection rate (DR, *NofI*), accuracy (E_s, E_r) and computational cost (*NFE*).

Experimental results that consider real and synthetic data provide evidence on the remarkable performance of HSRANSAC algorithm in comparison to the other methods. Additionally, in order to demonstrate the performance of the mentioned approach in a real engineering application, it has been employed to solve the problem of position estimation in a humanoid robot, with good results in terms of efficiency and accuracy.

References

1. H.-Y. Shum R. Szeliski, "Creating full view panoramic image mosaics and texture-mapped models," *Proceedings of the Computer Graphics, SIGGRAPH'97*, pp. 251–258, 1997.
2. R. Chung Y. He, "Image mosaicking for polyhedral scene and in particular singly visible surfaces," *Pattern Recognition*, vol. 31, no. 2008, pp. 1200–1213, 2008.
3. D.G. Lowe M. Brown, "Automatic panoramic image stitching using invariant features," *International Journal of Computer Vision*, vol. 2007, pp. 59–73, 2007.
4. M. Gökmen A. Akyol, "Super-resolution reconstruction of faces by enhanced global models of shape and texture," *Pattern Recognition*, vol. 45, no. 2012, pp. 4103–4116, 2012.
5. H.T. He, X. Fan, J.P. Zhang H. Huang, "Super-resolution of human face image using canonical correlation analysis," *Pattern Recognition*, vol. 43, no. 2010, pp. 2532–2543, 2010.
6. S.G. Gong K. Jia, "Hallucinating multiple occluded face images of different resolutions," *Pattern Recognition Letters*, vol. 27, no. 2006, pp. 1768–1775, 2006.
7. G. Bueno, E. Bermejo, R. Sukthankar O. Deniz, "Fast and accurate global motion compensation," *Pattern Recognition*, vol. 44, no. 12, pp. 2887–2901, 2011.
8. C. Sagues E. Montijano, "Distributed multi-camera visual mapping using topological maps of planar regions," *Pattern Recognition*, vol. 44, no. 7, pp. 1528–1539, 2011.
9. R. Chung, L. Jin J. Su, "Homography-based partitionng of curved surface for stereo correspondence establishment," *Pattern Recognition Letters*, vol. 28, no. 12, pp. 1459–1471, 2007.
10. C.H. Morimoto T.T. Santos, "Multiple camera people detection and tracking using support integration," *Pattern Recognition Letters*, vol. 32, no. 1, pp. 47–55, 2011.
11. J.P. Ostrowski H. Zhang, "Visual Motion Planning for Mobile Robots," *IEEE Transactions on Robotics and Automation*, vol. 18, no. 2, pp. 199–208, 2002.
12. J.J. Guerrero C. Sagüés, "Visual correction for mobile robot homing," *Robotics and Autonomous Systems*, vol. 50, no. 1, pp. 41–49, 2005.
13. J.J. Guerrero, C. Sagüés G. López-Nicolás, "Visual control of vehicles using two-view geometry," *Mechatronics*, vol. 20, no. 2, pp. 315–325, 2010.
14. M. A. Fischler and R. C. Bolles, "Random Sample Consensus: A Paradigm for Model Fitting with Applications to Image Analysis and Automated Cartography," *Comm. ACM*, vol. 24, no. 1981, pp. 381–395, 1981.
15. Adrian Hilton Evren İmre, "Order Statistics of RANSAC and Their Practical Application," *International Journal of Computer Vision*, vol. 2014, no. 1, 2014.
16. Qiushi Zhao, Wei Bu Xiangqian Wu, "A SIFT-based contactless palmprint verification approach using iterative RANSAC and local palmprint descriptors," *Pattern Recognition*, vol. 47, no. 10, pp. 3314–3326, 2014.
17. Yi Cui, Yexin Wang, Liu Liu, He Gao Fuqiang Zhou, "Accurate and robust estimation of camera parameters using RANSAC," *Optics and Lasers in Engineering*, vol. 51, no. 3, pp. 197–212, 2013.
18. O. Chum J. Matas, "Randomized RANSAC with Td, d test," *Image and Vision Computing*, vol. 22, no. 2004, pp. 837–842, 2004.
19. Shang-Hong Lai Chia-Ming Cheng, "A consensus sampling technique for fast and robust model fitting," *Pattern Recognition*, vol. 42, no. 2009, pp. 1318–1329, 2009.
20. P. H. S. Torr, "MLESAC: A New Robust Estimator with Application to Estimating Image Geometry," *Computer Vision and Image Understanding*, vol. 78, no. 2000, pp. 138–156, 2000.
21. Stuart B. Heinrich, "Efficient and robust model fitting with unknown noise scale," *Image and Vision Computing*, vol. 31, no. 2013, pp. 735–747, 2013.
22. P. Meer R. Subbarao, "Beyond RANSAC: user independent robust regression," *Proceedings IEEE International Conference on Computer Vision and Pattern Recognition Workshop*, vol. 1, no. 2006, p. 101, 2006.

23. D. Suter H. Wang, "Robust adaptive-scale parametric model estimation for computer vision," *IEEE Transactions on Pattern Analysis and Machine Intelligence*, vol. 26, no. 11, pp. 1459–1474, 2004.

24. Li Ming Li Jun-hua, "An analysis on convergence and convergence rate estimate of elitist genetic algorithms in noisy environments," *Optik - International Journal for Light and Electron Optics*, vol. 124, no. 24, pp. 6780–6785, 2013.

25. Renato A. Krohling, Mauro Campos Eduardo Mendel, "Swarm algorithms with chaotic jumps applied to noisy optimization problems," *Information Sciences*, vol. 181, no. 20, pp. 4494–4514, 2011.

26. Ling Wang, Bo Liu Hui Pan, "Particle swarm optimization for function optimization in noisy environment," *Applied Mathematics and Computation*, vol. 181, no. 2006, pp. 908–919, 2006.

27. Hans-Georg Beyer, "Evolutionary algorithms in noisy environments: theoretical issues and guidelines for practice," *Comput. Methods Appl. Mech. Engrg*, vol. 186, no. 2000, pp. 239–267, 2000.

28. J.H. Kim, G.V. Loganathan Z.W. Geem, "A new heuristic optimization algorithm: harmony search," *Simulations*, vol. 76, no. 2001, pp. 60–68, 2001.

29. M. Fesanghary, E. Damangir M. Mahdavi, "An improved harmony search algorithm for solving optimization problems," *Appl. Math. Comput.*, vol. 188, no. 2007, pp. 1567–1579, 2007.

30. M. Mahdavi M.G.H. Omran, "Global-best harmony search," *Appl. Math. Comput.*, vol. 198, no. 2008, pp. 643–656, 2008.

31. Z.W. Geem K.S. Lee, "A new meta-heuristic algorithm for continuous engineering optimization, harmony search theory and practice," *Comput. Methods Appl. Mech. Eng.*, vol. 194, no. 2005, pp. 3902–3933, 2005.

32. Z.W. Geem, S. H Lee, K.-W. Bae K.S. Lee, "The harmony search heuristic algorithm for discrete structural optimization," *Eng. Optim.*, vol. 37, no. 2005, pp. 663–684, 2005.

33. Z.W. Geem K.S. Lee, "A new structural optimization method based on the harmony search algorithm," *Comput. Struct.*, vol. 82, no. 2004, pp. 781–798, 2004.

34. Z.W. Geem, E.S. Kim J.H. Kim, "Parameter estimation of the nonlinear Muskingum model using harmony search," *J. Am. Water Resour. Assoc.*, vol. 37, no. 2001, pp. 1131–1138, 2001.

35. Z.W. Geem, "Optimal cost design of water distribution networks using harmony search," *Eng. Optim.*, vol. 38, no. 2006, pp. 259–280, 2006.

36. K.S. Lee, Y.J. Park Z.W. Geem, "Application of harmony search to vehicle routing," *Am. J. Appl. Sci.*, vol. 2, no. 2005, pp. 1552–1557, 2005.

37. Erik Cuevas, Gonzalo Pajares, Daniel Zaldivar, Marco Perez-Cisneros Diego Oliva, "Multilevel Thresholding Segmentation Based on Harmony Search Optimization," *Journal of Applied Mathematics*, vol. 2013, no. 1, pp. 1–24, 2013.

38. Noé Ortega-Sánchez, Daniel Zaldivar, Marco Pérez-Cisneros Erik Cuevas, "Circle Detection by Harmony Search Optimization," *Journal of Intelligent & Robotic Systems*, vol. 66, no. 3, pp. 359–376, 2012.

39. and S. Talataharib A. Kaveha, "Particle swarm optimizer, ant colony strategy and harmony search scheme hybridized for optimization of truss structures," *Computers & Structures*, vol. 87, no. 5–6, pp. 267-283, 2009.

40. and Zong Woo Geem Sungho Mun, "Determination of individual sound power levels of noise sources using a harmony search algorithm," *International Journal of Industrial Ergonomics*, vol. 39, no. 2, pp. 366–370, 2009.

41. and Zong Woo Geem Sungho Mun, "Determination of viscoelastic and damage properties of hot mix asphalt concrete using a harmony search algorithm," *Mechanics of Materials*, vol. 41, no. 3, pp. 339–353, 2009.

42. P.N. Suganthan, J.J. Liang, M. Fatih Tasgetiren Quan-Ke Pan, "A local-best harmony search algorithm with dynamic sub-harmony memories for lot-streaming flow shop scheduling problem," *Expert Systems with Applications*, vol. 38, no. 2011, pp. 3252–3259, 2011.

43. P.N. Suganthan, M. Fatih Tasgetiren, J.J. Liang Quan-Ke Pan, "A self-adaptive global best harmony search algorithm for continuous optimization problems," *Applied Mathematics and Computation*, vol. 216, no. 2010, pp. 830–848, 2010.
44. A. Ess, T. Tuytelaars, L.V. Gool H. Bay, "Surf: speeded up robust features," *Computer Vision and Image Understanding (CVIU)*, vol. 110, no. 3, pp. 346–359, 2008.
45. M. Stephens C. Harris, "A combined corner and edge detector," *Proceedings of the 4th Alvey Vision Conference*, vol. 1, no. 1, pp. 147–151, 1988.
46. P. Meer, "Robust Techniques in Computer Vision," *Emerging Topics in Computer Vision, Ed. by G. Medioni and S. B. Kang (Prentice Hall, Boston, 2004)*, vol. 1, no. 2004, pp. 107–190, 2004.
47. Quan-ke Pan, Jun-qing Li Jing Chen, "Harmony search algorithm with dynamic control parameters," *Applied Mathematics and Computation*, vol. 219, no. 2, pp. 592–604, 2012.
48. James Kennedy Maurice Clerc, "The Particle Swarm-Explosion, Stability, and Convergence in a Multidimensional Complex Space," *IEEE Transactions on Evolutionary Computation*, vol. 6, no. 1, pp. 58–73, 2002.
49. Pooya Mirzabeygi, Masoud Shariat Panahi Behrooz Ostadmohammadi Arani, "An improved PSO algorithm with a territorial diversity-preserving scheme and enhanced exploration-exploitation balance," *Swarm and Evolutionary Computation*, vol. 11, no. 2013, pp. 1–15, 2013.
50. Jung-Han Kim Ah-Lam Lee, "3-Dimensional pose sensor algorithm for humanoid robot," *Control Engineering Practice*, vol. 18, no. 10, pp. 1173–1182, 2010.
51. Kamal Jamshidi, Amir Hasan Monadjemi, Hafez Eslami Hamed Shahbazi, "Biologically inspired layered learning in humanoid robots," *Knowledge-Based Systems*, vol. 57, no. 2014, pp. 8–27, 2014.
52. G. Schmidt J.F. Seara, "Intelligent gaze control for vision-guided humanoid walking: methodological aspects," *Robotics and Autonomous Systems*, vol. 48, no. 4, pp. 231–248, 2004.

Chapter 10
Filter Design

Abstract System identification is a complex optimization problem which has recently attracted the attention in the field of science and engineering. In particular, the use of infinite impulse response (IIR) models for identification is preferred over their equivalent FIR (finite impulse response) models since the former yield more accurate models of physical plants for real world applications. However, IIR structures tend to produce multimodal error surfaces for which their cost functions are significantly difficult to minimize. Evolutionary computation techniques (ECT) are used to estimate the solution to complex optimization problems. They are often designed to meet the requirements of particular problems because no single optimization algorithm can solve all problems competitively. Therefore, when new algorithms are proposed, their relative efficacies must be appropriately evaluated. Several comparisons among ECT have been reported in the literature. Nevertheless, they suffer from one limitation: their conclusions are based on the performance of popular evolutionary approaches over a set of synthetic functions with exact solutions and well-known behaviors, without considering the application context neither including recent developments. This study presents the comparison of various evolutionary computation optimization techniques applied to IIR model identification. In the comparison, special attention is paid to recently developed algorithms such as Cuckoo Search and Flower Pollination Algorithm, including also popular approaches. Results over several models are presented and statistically validated.

10.1 Introduction

System identification is a complex optimization problem which has recently attracted the attention in the field of science and engineering. System identification is important in the disciplines of control systems [1], communication [2], signal processing [3] and image processing [4].

In a system identification configuration, an optimization algorithm attempts to iteratively determine the adaptive model parameters to get an optimal model for an unknown plant by minimizing some error function between the output of the

© Springer International Publishing AG 2017   205
E. Cuevas et al., *Evolutionary Computation Techniques:*
A Comparative Perspective, Studies in Computational Intelligence 686,
DOI 10.1007/978-3-319-51109-2_10

candidate model and the output of the plant. The optimal model or solution is attained when such error function is effectively reduced. The adequacy of the estimated model depends on the adaptive model structure, the optimization algorithm, and also the characteristic and quality of the input-output data [5].

Systems or plants can be better modeled through infinite impulse response (IIR) models because they emulate physical plants more accurately than their equivalent FIR (finite impulse response) models [6]. In addition, IIR models are typically capable of meeting performance specifications using fewer model parameters. However, IIR structures tend to produce multimodal error surfaces whose cost functions are significantly difficult to minimize [7]. Hence, in order to identify IIR models, a practical, efficient, and robust global optimization algorithm is necessary to minimize the multimodal error function.

Traditionally, the least mean square (LMS) technique and its variants [8] have been extensively used as optimization tools for IIR model identification. The wide acceptance of such gradient based optimization techniques is due to the low complexity and simplicity of implementation. However, the error surface for the IIR model is mostly multi-modal with respect to the filter coefficients. This may result in leading traditional gradient-descent approaches into local optima [9].

The difficulties associated with the use of gradient based optimization methods for solving several engineering problems have contributed to the development of alternative solutions. Evolutionary computation techniques (ECT) such as the particle swarm optimization (PSO) [10], artificial bee colony (ABC) [11], electromagnetism-like method (EM) [12], Cuckoo Search (CS) [13] and Flower Pollination Algorithm (FPA) [14] have received much attention regarding their potential as global optimization methods in real-world applications. Inspired by the evolution process and survival of the fittest in the biological world, ECT are search methods that are different from traditional optimization methods. They are based on a collective learning process within a population of candidate solutions. The population in ECT is usually arbitrarily initialized, and each iteration (also called a generation) evolves towards better and better solution regions by means of randomized processes where several operators are applied to each candidate solution. ECT have been applied to many engineering optimization problems and have proven to be effective for solving some specific problems, including multimodal optimization, dynamic optimization, noisy optimization, multi-objective optimization, and so on [15–17]. Hence, they are becoming increasingly popular tools to solve various hard optimization problems.

As an alternative to gradient based techniques, the problem of IIR modelling has also been handled through evolutionary computation techniques. In general, they have demonstrated to yield better results than those based on gradient algorithms with respect to accuracy and robustness [9]. Such approaches have produced several robust IIR identification systems by using different evolutionary computation techniques such as PSO [18], ABC [19], EM [20] and CS [21], whose results have been individually reported.

ECT are often designed to meet the requirements of particular problems because no single optimization algorithm can solve all problems competitively [22]. Therefore, when new alternative algorithms are proposed, their relative efficiency must be appropriately evaluated. Many efforts [23–25] have also been devoted to compare ECT to each other. Typically, such comparisons have been based on synthetic numerical benchmark problems with most studies verifying if one algorithm outperforms others over a given set of benchmarks functions overlooking any statistical test. However, few comparative studies of various ECT considering the application context are available in the literature. Therefore, it is very important to discuss and compare the performance of ECT methods from an application point of view.

This paper presents the comparison of various evolutionary computation optimization techniques that are applied to IIR model identification. In the comparison, special attention is paid to recently developed algorithms such as the Cuckoo Search (CS) and the Flower Pollination Algorithm (FPA), including also popular approaches as the Particle Swarm Optimization (PSO), the Artificial Bee Colony (ABC) optimization and the Electromagnetism-Like Optimization (EM) algorithm. Results over several models with different ranges of complexity are presented and validated within a statistically significant framework.

The rest of this paper is organized as follows: Sect. 10.2 presents a review of the evolutionary computation techniques that are employed in the comparison whereas Sect. 10.3 discusses on the IIR system identification problem. In Sect. 10.4 all experimental results are depicted with some concluding remarks being drawn in Sect. 10.5.

10.2 Evolutionary Computation Techniques (ECT)

In the real world, many optimization problems can be considered as black box challenges. Often, few information is available about an optimization problem itself unless the information emerging from function evaluations. In the worst case, nothing is known about the characteristics of the fitness function, e.g., whether it is unimodal or multimodal.

On the other hand, ECT are used to estimate the solution to complex optimization problems since they adapt easily to black-box formulations and extremely ill-behaved functions. ECT are based on a collective learning process within a population of candidate solutions. The population in ECT is usually arbitrarily initialized while each iteration (also called a generation) evolves towards better solution regions by means of randomized processes with several operators being applied to each candidate solution. ECT have been applied to many engineering optimization problems ensuring an effective solution for some specific problems, including multimodal optimization, dynamic optimization, noisy optimization, multi-objective optimization, and others [15–17].

Therefore, ECT are becoming increasingly popular tools to solve various hard optimization problems. This section presents a brief description of five evolutionary computation techniques: Swarm Optimization (PSO), Artificial Bee Colony (ABC) Optimization and Electromagnetism-Like Optimization (EM), Cuckoo Search (CS) and Flower Pollination Algorithm (FPA), which have been all employed in our comparative study.

10.2.1   Particle Swarm Optimization (PSO)

PSO, proposed by Kennedy and Eberhart in 1995 [10], is a population-based stochastic optimization technique that is inspired on the social behavior of bird flocking or fish schooling. The algorithm searches for the optimum using a group or swarm formed by possible solutions of the problem, which are called particles. From the implementation point of view, in the PSO operation, a population $\mathbf{P}^k$ ($\{\mathbf{p}_1^k, \mathbf{p}_2^k, \ldots, \mathbf{p}_N^k\}$) of N particles (individuals) evolves from the initial point ($k = 0$) to a total *gen* number iterations ($k = gen$). Each particle $\mathbf{p}_i^k$ ($i \in [1, \ldots, N]$) represents a d-dimensional vector $\{p_{i,1}^k, p_{i,2}^k, \ldots, p_{i,d}^k\}$ where each dimension corresponds to a decision variable of the optimization problem at hand. The quality of each particle $\mathbf{p}_i^k$ (candidate solution) is evaluated by using an objective function $f(\mathbf{p}_i^k)$ whose final result represents the fitness value of $\mathbf{p}_i^k$. During the evolution process, the best global position $\mathbf{g}$ ($g_1, g_2, \ldots g_d$) seen so-far is stored with the best position $\mathbf{p}_i^*$ ($p_{i,1}^*, p_{i,2}^*, \ldots, p_{i,d}^*$) being reached by each particle. Such positions are computed by considering a minimization problem as follows:

$$\mathbf{g} = \underset{i \in \{1,2,\ldots,N\}, a \in \{1,2,\ldots,k\}}{\arg\min} (f(\mathbf{p}_i^a)) \quad \mathbf{p}_i^* = \underset{a \in \{1,2,\ldots,k\}}{\arg\min} (f(\mathbf{p}_i^a)). \quad (10.1)$$

In this work, the modified PSO version proposed by Lin in [26] has been implemented. Under such approach, the new position $\mathbf{p}_i^{k+1}$ of each particle $\mathbf{p}_i^k$ is calculated by using the following equations:

$$\begin{aligned} v_{i,j}^{k+1} &= w \cdot v_{i,j}^k + c_1 \cdot r_1 \cdot (p_{i,j}^* - p_{i,j}^k) + c_2 \cdot r_2 \cdot (g_j - p_{i,j}^k); \\ p_{i,j}^{k+1} &= p_{i,j}^k + v_{i,j}^{k+1}; \end{aligned} \quad (10.2)$$

where w is called the inertia weight that controls the impact of the current velocity on the updated velocity ($i \in [1, \ldots, N]$, $j \in [1, \ldots, d]$). c_1 and c_2 are the positive acceleration coefficients that rule the movement of each particle towards the positions $\mathbf{g}$ and $\mathbf{p}_i^*$, respectively. r_1 and r_2 are uniformly distributed random numbers that are chosen within the interval [0, 1].

10.2.2 Artificial Bee Colony (ABC)

The artificial bee colony (ABC) algorithm, proposed by Karaboga [11], is an ECT inspired by the intelligent foraging behavior of a honeybee swarm. In the ABC operation, a population $\mathbf{L}^k$ ($\{\mathbf{l}_1^k, \mathbf{l}_2^k, \ldots, \mathbf{l}_N^k\}$) of N food locations (individuals) is evolved from the initial point ($k = 0$) to a total *gen* number iterations ($k = gen$). Each food location $\mathbf{l}_i^k$ ($i \in [1, \ldots, N]$) represents a *d*-dimensional vector $\{l_{i,1}^k, l_{i,2}^k, \ldots, l_{i,d}^k\}$ where each dimension corresponds to a decision variable of the optimization problem to be solved. After initialization, an objective function evaluates each food location to assess whether it represent an acceptable solution (nectar-amount) or not. Guided by the values of such an objective function, the candidate solution $\mathbf{l}_i^k$ is evolved through different ABC operations (honey bee types). In the main operator, each food location $\mathbf{l}_i^k$ generates a new food source $\mathbf{t}_i$ in the neighborhood of its present position as follows:

$$\mathbf{t}_i = \mathbf{l}_i^k + \phi(\mathbf{l}_i^k - \mathbf{l}_r^k), \quad i, r \in (1, 2, \ldots, N) \tag{10.3}$$

where $\mathbf{l}_r^k$ is a randomly chosen food location, satisfying the condition $r \neq i$. The scale factor ϕ is a random number between $[-1, 1]$. Once a new solution $\mathbf{t}_i$ is generated, a fitness value representing the profitability associated with a particular solution $fit(\mathbf{l}_i^k)$ is calculated. The fitness value for a minimization problem can be assigned to a candidate solution $\mathbf{l}_i^k$ by the following expression:

$$fit(\mathbf{l}_i^k) = \begin{cases} \frac{1}{1+f(\mathbf{l}_i^k)} & \text{if } f(\mathbf{l}_i^k) \geq 0 \\ 1 + abs(f(\mathbf{l}_i^k)) & \text{if } f(\mathbf{l}_i^k) < 0 \end{cases} \tag{10.4}$$

where $f(\cdot)$ represents the objective function to be minimized. Once the fitness values are calculated, a greedy selection process is applied between $\mathbf{t}_i$ and $\mathbf{l}_i^k$. If $fit(\mathbf{t}_i)$ is better than $fit(\mathbf{l}_i^k)$, then the candidate solution $\mathbf{l}_i^k$ is replaced by $\mathbf{t}_i$; otherwise, $\mathbf{l}_i^k$ remains.

10.2.3 Electromagnetism-like (EM) Algorithm

The EM algorithm, proposed by İlker-Birbil and Shu-Cherng [12], is a simple and population-based search algorithm which has been inspired by the electro-magnetism phenomenon. In EM, individuals emulate charged particles which interact to each other based on the electro-magnetism laws of repulsion and attraction. The method utilizes N, *d*-dimensional points $\mathbf{x}_i^k$, $i = 1, 2, \ldots, N$ where each point $\mathbf{x}_i^k$ is a d-dimensional vector containing the parameter values to be optimized ($\mathbf{x}_i^k = \{x_{i,1}^k, \ldots, x_{i,d}^k\}$) whereas k denotes the iteration (or generation) number. The initial population $\mathbf{X}^k = \{\mathbf{x}_1^k, \mathbf{x}_2^k, \ldots, \mathbf{x}_N^k\}$ (being $k = 0$), is taken from uniformly distributed samples of the search space. We denote the population set at the k th generation by $\mathbf{X}^k$, because members of $\mathbf{X}^k$ change with k. After the initialization of $\mathbf{X}^0$, EM continues its iterative process until a stopping condition (e.g. the maximum number of generations, $k = gen$) is met. An iteration of EM consists of three steps. In the first step each point in $\mathbf{X}^k$ moves to a different location by using the attraction-repulsion mechanism of the electromagnetism theory. In the second step, points moved by the electromagnetism principle are further moved locally by a local search procedure. Finally, in the third step, in order to generate the new population $\mathbf{X}^{k+1}$, a greedy selection process selects best points between those produced by the local search procedure and the originals. Both the attraction-repulsion mechanism and the local search in EM are responsible for driving the members $\mathbf{x}_i^k$ of $\mathbf{X}^k$ to the proximity of the global optimum.

10.2.4 Cuckoo Search (CS) Method

CS is one of the latest nature-inspired algorithms that has been developed by Yang and Deb [13]. CS is based on the brood parasitism of some cuckoo species. In addition, this algorithm is enhanced by the so-called Lévy flights [46], rather than by simple isotropic random walks. From the implementation point of view of the CS operation, a population $\mathbf{E}^k$ ($\{\mathbf{e}_1^k, \mathbf{e}_2^k, \ldots, \mathbf{e}_N^k\}$) of N eggs (individuals) is evolved from the initial point ($k = 0$) to a total gen number iterations ($k = 2 \cdot gen$). Each egg $\mathbf{e}_i^k$ ($i \in [1, \ldots, N]$) represents a d-dimensional vector $\left\{e_{i,1}^k, e_{i,2}^k, \ldots, e_{i,d}^k\right\}$ where each dimension corresponds to a decision variable of the optimization problem to be solved. The quality of each egg $\mathbf{e}_i^k$ (candidate solution) is evaluated by using an objective function $f\left(\mathbf{e}_i^k\right)$ whose final result represents the fitness value of $\mathbf{e}_i^k$. Three different operators define the evolution process of CS: (A) Lévy flight, (B) the Replace of Nests operator for constructing new solutions and (C) the Elitist selection strategy.

(A) The Lévy flight

One of the most powerful features of cuckoo search is the use of Lévy flights to generate new candidate solutions (eggs). Under this approach, a new candidate solution $\mathbf{e}_i^{k+1}$ ($i \in [1, \ldots, N]$) is produced by perturbing the current $\mathbf{e}_i^k$ with a change of position $\mathbf{c}_i$. In order to obtain $\mathbf{c}_i$, a random step $\mathbf{s}_i$ is generated by a symmetric

Lévy distribution. For producing s_i, the Mantegna's algorithm [47] is employed as follows:

$$s_i = \frac{\mathbf{u}}{|\mathbf{v}|^{1/\beta}}, \tag{10.5}$$

where $\mathbf{u}$ $(\{u_1, \ldots, u_d\})$ and $\mathbf{v}$ $(\{v_1, \ldots, v_d\})$ are n-dimensional vectors and $\beta = 3/2$. Each element of $\mathbf{u}$ and $\mathbf{v}$ is calculated by considering the following normal distributions:

$$u \sim N(0, \sigma_u^2), \quad v \sim N(0, \sigma_v^2),$$
$$\sigma_u = \left(\frac{\Gamma(1+\beta) \cdot \sin(\pi \cdot \beta/2)}{\Gamma((1+\beta)/2) \cdot \beta \cdot 2^{(\beta-1)/2}} \right)^{1/\beta}, \quad \sigma_v = 1, \tag{10.6}$$

where $\Gamma(\cdot)$ represents the Gamma distribution. Once s_i has been calculated, the required change of position $\mathbf{c}_i$ is computed as follows:

$$\mathbf{c}_i = 0.01 \cdot s_i \oplus (\mathbf{e}_i^k - \mathbf{e}^{best}), \tag{10.7}$$

where the product $\oplus$ denotes entry-wise multiplications whereas $\mathbf{e}^{best}$ is the best solution (egg) seen so far in terms of its fitness value. Finally, the new candidate solution $\mathbf{e}_i^{k+1}$ is calculated by using:

$$\mathbf{e}_i^{k+1} = \mathbf{e}_i^k + \mathbf{c}_i \tag{10.8}$$

(B) Replace some nests by constructing new solutions

Under this operation, a set of individuals (eggs) are probabilistically selected and replaced with a new value. Each individual $\mathbf{e}_i^k$ ($i \in [1, \ldots, N]$) can be selected with a probability $p_a \in [0, 1]$. In order to implement this operation, a uniform random number r_1 is generated within the range $[0, 1]$. If r_1 is less than p_a, the individual $\mathbf{e}_i^k$ is selected and modified according to Eq. 10.5; otherwise $\mathbf{e}_i^k$ remains with no change. This operation can be resumed by the following model:

$$\mathbf{e}_i^{k+1} = \begin{cases} \mathbf{e}_i^k + \text{rand} \cdot (\mathbf{e}_j^k - \mathbf{e}_h^k) & \text{with probability } p_a \\ \mathbf{e}_i^k & \text{with probability } (1 - p_a) \end{cases} \tag{10.9}$$

where *rand* is a random number normally distributed whereas j and h are random integers from 1 to N.

(C) The Elitist Selection Strategy

After producing $\mathbf{e}_i^{k+1}$ either by the operator A or the operator B, it must be compared with its past value $\mathbf{e}_i^k$. If the fitness value of $\mathbf{e}_i^{k+1}$ is better than $\mathbf{e}_i^k$, then $\mathbf{e}_i^{k+1}$ is accepted as the final solution; otherwise, $\mathbf{e}_i^k$ is retained. This procedure can be resumed by the following statement:

$$\mathbf{e}_i^{k+1} = \begin{cases} \mathbf{e}_i^{k+1} & \text{if } f(\mathbf{e}_i^{k+1}) < f(\mathbf{e}_i^k) \\ \mathbf{e}_i^k & \text{otherwise} \end{cases} \tag{10.10}$$

The elitist selection strategy denotes that only high-quality eggs (best solutions near to the optimal value) which are the most similar to the host bird's eggs have the opportunity to develop (next generation) and become mature cuckoos.

10.2.5 Flower Pollination Algorithm (FPA)

The flower pollination algorithm (FPA), proposed by Yang [14], is an ECT inspired by the pollination process of flowers. In FPA, individuals emulate a set of flowers or pollen gametes which behaves based on biological laws of the pollination process. From the implementation point of view, in the FPA operation, a population $\mathbf{F}^k(\{\mathbf{f}_1^k, \mathbf{f}_2^k, \ldots, \mathbf{f}_N^k\})$ of N flower positions (individuals) is evolved from the initial point $(k = 0)$ to a total gen number iterations $(k = gen)$. Each flower $\mathbf{f}_i^k$ $(i \in [1, \ldots, N])$ represents a d-dimensional vector $\{f_{i,1}^k, f_{i,2}^k, \ldots, f_{i,d}^k\}$ where each dimension corresponds to a decision variable of the optimization problem to be solved. In FPA, a new population $\mathbf{F}^{k+1}$ is produced by considering two operators: local and global pollination. A probabilistic global pollination factor p is associated with such operators. In order to decide which operator should be applied to each current flower position $\mathbf{f}_i^k$, a uniform random number r_p is generated within the range $[0, 1]$. If r_p is less than p, the global pollination operator is applied to $\mathbf{f}_i^k$. Otherwise, the local pollination operator is considered.

Global pollination operator.

Under this operator, the original position $\mathbf{f}_i^k$ is displaced to a new position $\mathbf{f}_i^{k+1}$ according to the following model:

$$\mathbf{f}_i^{k+1} = \mathbf{f}_i^k + s_i \cdot (\mathbf{f}_i^k - \mathbf{g}), \tag{10.11}$$

where $\mathbf{g}$ is the global best position seen so-far whereas s_i controls the length of the displacement. The s_i value is generated by a symmetric Lévy distribution according to Eqs. 10.5–10.6.

Local pollination operator.

In the local pollination operator, the current position $\mathbf{f}_i^k$ is perturbed to a new position $\mathbf{f}_i^{k+1}$ as follows:

$$\mathbf{f}_i^{k+1} = \mathbf{f}_i^k + \varepsilon \cdot (\mathbf{f}_j^k - \mathbf{f}_h^k); \quad i,j,h \in (1,2,\ldots,N), \tag{10.12}$$

where $\mathbf{f}_j^k$ and $\mathbf{f}_h^k$ are two randomly chosen flower positions, satisfying the condition $j \neq h \neq i$. The scale factor ε is a random number between $[-1,1]$.

10.3   IIR Model Identification (Problem Formulation)

System identification is the mathematical representation of an unknown system by using only input–output data. In a system identification configuration, an optimization algorithm attempts to iteratively determine the adaptive model parameters to get an optimal model for the unknown plant based on minimizing some error function between the output of the candidate model and the actual output of the plant.

The use of infinite impulse response (IIR) models for identification is preferred over their equivalent FIR (finite impulse response) models since the former produce more accurate models of physical plants for real world applications [6]. In addition, IIR models are typically capable of meeting performance specifications using fewer model parameters. Figure 10.1 represents an IIR identification model of any arbitrary system.

An IIR system can be represented by the transfer function:

$$\frac{Y(z)}{X(z)} = \frac{b_0 + b_1 z^{-1} + b_2 z^{-2} + \cdots + b_m z^{-m}}{1 + a_1 z^{-1} + a_2 z^{-2} + \cdots + a_n z^{-n}}, \tag{10.13}$$

where m and n are the number of numerator and denominator coefficients of the transfer function respectively and, a_i and b_j are the pole and zero parameters of the IIR model ($i \in [1,\ldots,n]$, $j \in [1,\ldots,m]$). Equation 10.13 can be written as difference equation of the form:

$$y(t) = \sum_{i=1}^{n} a_i \cdot y(t-i) + \sum_{j=0}^{m} b_j \cdot x(t-j), \tag{10.14}$$

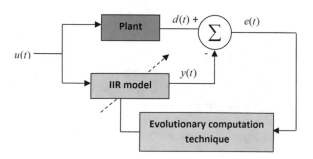

Fig. 10.1 Adaptive IIR model for system identification

where $u(t)$ and $y(t)$ represent the t th input and output of the system, respectively. Therefore, the set of unknown parameters that models the IIR system is represented by $\theta = \{a_1, \ldots, a_n, b_0, \ldots, b_m\}$. Considering that the number of unknown parameters of θ is $(n + m + 1)$, the search space $\mathbf{S}$ of feasible values for θ is $\Re^{(n+m+1)}$.

According to the block diagram of Fig. 10.1, the output of the plant is d (t) whereas the output of the IIR filter is $y(t)$. The output difference between the actual system and its model yields the error $e(t) = d(t)-y(t)$. Hence, the problem of IIR model identification can be considered as a minimization problem of the function $f(\theta)$ stated as the following:

$$f(\theta) = \frac{1}{W} \sum_{t=1}^{W} (d(t) - y(t))^2, \qquad (10.15)$$

where W is the number of samples used in the simulation.

The aim is to minimize the cost function $f(\theta)$ by adjusting θ. The optimal model θ^* or solution is attained when the error function $f(\theta)$ reaches its minimum value, as follows:

$$\theta^* = \arg \min_{\theta \in \mathbf{S}} (f(\theta)), \qquad (10.16)$$

10.4 Experimental Results

In the comparison study, a comprehensive set of experiments has been used to test the performance of each evolutionary computation technique. The set considers the use of IIR models with different orders. Such experimental set has been carefully selected to assure compatibility between similar works reported in the literature [18–21]. In the comparison, five ETC have been considered: PSO, ABC, EM, CS and FPA.

The parameter setting for each evolutionary computation algorithm that is used in the comparison is described as follows:

1. PSO: The parameters are set to $c_1 = 2$, $c_2 = 2$; besides, the weight factor decreases linearly from 0.9 to 0.2 [18].
2. ABC: The algorithm has been implemented using the guidelines provided by its own reference [19], using the parameter *limit* = 100.
3. EM: particle number = 50, $\delta = 0.001$, *LISTER* = 4, *MaxIter* = 300. Such values, according to [12, 20] represent the best possible configuration.
4. CS: According to [13, 21], the parameters are set to $p_a = 0.25$ and the number of generations *gen* = 500.
5. FPA: the probabilistic global pollination factor p is set to 0.8. Under such value, the algorithm presents the best performance according to [14].

For all algorithms, the population size has been set to 25 ($N = 25$) whereas the maximum iteration number has been configured to 3000 generations ($gen = 3000$).

The results are divided into two sections. In the first set, the performance of each ETC for each identification experiment is presented. In the second set, the results are analyzed from a statistical point of view by using the Wilcoxon test.

10.4.1 IIR Model Identification Results

The results are reported considering three experiments that include, (1) a second-order plant with a first-order IIR model; (2) a second-order plant with a second-order IIR model; finally, (3) a high-order plant with a high-order model. Each case is discussed below.

(1) A plant with a second-order system and a first-order IIR model (first experiment)

In this experiment, each algorithm is applied to identify a second order plant through a first-order IIR model. Under such conditions, the unknown plant H_P and the IIR model H_M hold the following transfer functions:

$$H_P(z^{-1}) = \frac{0.05 - 0.4z^{-1}}{1 - 1.1314z^{-1} + 0.25z^{-2}}, \quad H_M(z^{-1}) = \frac{b}{1 - az^{-1}} \tag{10.17}$$

In the simulations, it has been considered a white sequence of 100 samples ($W = 100$) for the input $u(t)$. Since a reduced order model is employed to identify a plant of a superior order, $f(\theta)$ is multi-modal [19]. The error surface $f(\theta)$ is shown in Fig. 10.2.

The performance evaluation over 30 different executions is reported in Table 10.1 considering the following indexes: the best parameter values (ABP), the average $f(\theta)$ value (AV) and the standard deviation (SD). The best parameter values (ABP) report the best model parameters obtained during the 30 executions while the average $f(\theta)$ value (AV) indicates the average minimum value of $f(\theta)$, considering the same number of executions. Finally, the standard deviation (SD) reports the dispersion from the average $f(\theta)$ value regarding 30 executions.

According to Table 10.1, the CS algorithm provides better results than PSO, ABC and EM. In particular, the results show that CS maintains a considerable precision (the lowest AV value) and more robustness (smallest SD value). Nevertheless, the CS performance is similar to the FPA algorithm. On the other hand, the worst performance is reached by the PSO algorithm. Such a fact corresponds to its difficulty (premature convergence) to overcome local minima in multimodal functions.

Fig. 10.2 Multimodal error
surface $f(\theta)$ for the first
experiment: **a** tridimensional
figure and **b** contour

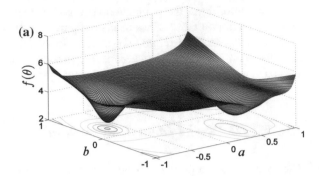

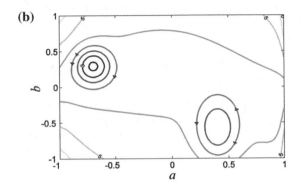

Table 10.1 Performance
results of the first experiment

Algorithms	ABP		AV	SD
	a	b		
PSO	0.9125	−0.3012	0.0284	0.0105
ABC	0.1420	−0.3525	0.0197	0.0015
EM	0.9034	0.3030	0.0165	0.0012
CS	0.9173	−0.2382	0.0101	3.118e-004
FPA	0.9364	−0.2001	0.0105	5.103e-004

(2) A plant with second-order system and second-order IIR model (second experiment)

In the second experiment, the performance for each algorithm is evaluated at the identification of a second order plant through a second-order IIR model. Therefore, the unknown plant H_P and the IIR model H_M hold the following transfer functions:

$$H_P(z^{-1}) = \frac{1}{1 - 1.4z^{-1} + 0.49z^{-2}}, \quad H_M(z^{-1}) = \frac{b}{1 + a_1 z^{-1} + a_2 z^{-2}} \quad (10.18)$$

For the simulations, the input $u(t)$, that is applied to the system and to the IIR model simultaneously, has been configured as a white sequence with 100 samples.

Table 10.2 Performance results of the second experiment

Algorithms	ABP			AV	SD
	a_1	a_2	b		
PSO	−1.4024	0.4925	0.9706	4.0035e-005	1.3970e-005
ABC	−1.2138	0.6850	0.2736	0.3584	0.1987
EM	−1.0301	0.4802	1.0091	3.9648e-005	8.7077e-005
CS	−1.400	0.4900	1.000	0.000	0.000
FPA	−1.400	0.4900	1.000	4.6246e-32	2.7360e-31

Since the order of the model H_M is equal to the order of the to-be-identified system H_P, only one global minimum exists in $f(\theta)$ [19]. The results of this experiment over 30 different executions are reported in Table 10.2.

The results in Table 10.2 show that PSO, ABC, EM, CS and FPA have similar values in their performance. The evidence shows that evolutionary algorithms maintain a similar average performance when they face unimodal low-dimensional functions [27, 28]. In particular, the test remarks that the small difference in performance is directly related to a better exploitation mechanism included in CS and FPA.

(3) A superior-order plant and a high-order model (third experiment)

Finally, the performance for each algorithm is evaluated at the identification of a superior-order plant through a high-order IIR model. Therefore, the unknown plant H_P and the IIR model H_M hold the following transfer functions:

$$H_P(z^{-1}) = \frac{1 - 0.4z^{-2} - 0.65z^{-4} + 0.26z^{-6}}{1 - 0.77z^{-2} - 0.8498z^{-4} + 0.6486z^{-6}},$$
$$H_M(z^{-1}) = \frac{b_0 + b_1z^{-1} + b_2z^{-2} + b_3z^{-3} + b_4z^{-4}}{1 + a_1z^{-1} + a_2z^{-2} + a_3z^{-3} + a_4z^{-4}}$$
(10.19)

Since the plant is a sixth-order system and the IIR model a fourth-order system, the error surface $f(\theta)$ is multimodal just as it is in the first experiment. A white sequence with 100 samples has been used as input. The results of this experiment over 30 different executions are reported in Tables 10.3 and 10.4. Table 10.3 presents the best parameter values (ABP) whereas Table 10.4 shows the average $f(\theta)$ value (AV) and its standard deviation (SD).

According to the AV and SD indexes in Table 10.4, the CS algorithm finds better results than PSO, ABC, EM and FPA. The results show that CS presents a better precision (AV value) and robustness (SD value). These results also indicate that CS, FPA and EM are able to identify the sixth order plant under different accuracy levels. On the other hand, PSO and ABC obtain sub-optimal solutions whose parameters weakly model the unknown system.

Table 10.3 The best parameter values (ABP) for the second experiment

Algorithms	ABP											
	a_1	a_2	a_3	a_4	b_0	b_1	b_2	b_3	b_4
PSO	0.3683	−0.704	0.2807	0.3818	0.9939	−0.660	−0.852	0.2275	−1.4990
ABC	−1.163	−0.635	−1.518	0.6923	0.5214	−1.270	0.3520	1.1816	−1.9411
EM	−0.495	−0.704	0.5656	−0.269	1.0335	−0.667	−0.468	0.6961	−0.0673
CS	0.9599	0.024	0.0368	−0.002	−0.237	0.0031	−0.357	0.0011	−0.5330
FPA	0.0328	−0.105	−0.024	−0.761	1.0171	0.0038	0.2374	0.0259	−0.3365

Table 10.4 The average $f(\theta)$ value (AV) and the standard deviation (SD)

Algorithms	AV	SD
PSO	5.8843	3.4812
ABC	7.3067	4.3194
EM	0.0140	0.0064
CS	6.7515e-004	4.1451e-004
FPA	0.0018	0.0020

10.4.2 Statistical Analysis

Recently, statistic tests have been used in several domains [29–34] to validate the performance of new approaches over an experimental data set. Therefore, in order to statistically validate the results, a non parametric statistical significance-proof which is known as the Wilcoxon's rank sum test for independent samples [35, 36], has been conducted over the "the average $f(\theta)$ value" (AV) data of Tables 10.1, 10.2 and 10.4 with an 5% significance level. The test has been conducted considering 30 different executions for each algorithm. Table 10.5 reports the p-values produced by Wilcoxon's test for the pair-wise comparison of the "the average $f(\theta)$ value" of four groups. Such groups are formed by CS versus. PSO, CS versus. ABC, CS versus. EM and CS versus. FPA. As a null hypothesis, it is assumed that there is no significant difference between averaged values of the two algorithms. The alternative hypothesis considers a significant difference between the AV values of both approaches.

For the case of PSO, ABC and EM, all p-values reported in Table 10.5 are less than 0.05 (5% significance level) which is a strong evidence against the null hypothesis. Therefore, such evidence indicates that CS results are statistically significant and that it has not occurred by coincidence (i.e. due to common noise contained in the process). On the other hand, since the p-values for the case of CS vs FPA are more than 0.05, there is not statistical difference between both. Therefore, it can be concluded that the CS algorithm is better than PSO, ABC and EM in the application of IIR modeling for system identification. However, CS presents the same performance than FPA and therefore there is not statistical evidence that CS surpasses the FPA algorithm.

Table 10.5 p-values produced by Wilcoxon's test comparing CS vs PSO, ABC, EM and FPA over the "The average $f(\theta)$ values (AV)" from Tables 10.1, 10.2 and 10.4

CS versus	PSO	ABC	EM	FPA
First experiment	6.5455e-13	8.4673e-13	3.8593e-08	0.7870
Second experiment	1.5346e-14	1.5346e-14	1.5346e-14	0.3313
Third experiment	6.5455e-13	1.5346e-14	4.3234e-13	0.1011

10.5 Conclusions

This paper presents a comparison study between five evolutionary algorithms for the IIR-based model identification. Under this research, the identification task is considered as an optimization problem. In the comparison, special attention is paid to recently developed algorithms such as the Cuckoo Search (CS) and the Flower Pollination Algorithm (FPA), also including popular approaches such as the Particle Swarm Optimization (PSO), the Artificial Bee Colony optimization (ABC) and the Electromagnetism-Like (EM) optimization algorithm.

The comparison has been experimentally evaluated over a test suite of three benchmark experiments that produce multimodal functions. The experiment results have demonstrated that CS outperforms PSO, ABC and EM in terms of both, the accuracy (AV values) and robustness (SD values), within a statistically significant framework (Wilcoxon test). However, there is not statistical evidence that CS surpasses the FPA performance.

The remarkable performance of CS and FPA is explained by two different features: (i) operators (such as Lévy flight) that allow a better exploration of the search space, increasing the capacity to find multiple optima; (ii) their exploitation operators that allow a better precision of previously-found solutions.

References

1. Xiaojun Zhou, Chunhua Yang, Weihua Gui. Nonlinear system identification and control using state transition algorithm, Applied Mathematics and Computation, 226, (2014), 169–179.
2. Mouayad Albaghdadia, Bruce Brileyb, Martha Evens, Event storm detection and identification in communication systems, Reliability Engineering and System Safety 91 (2006) 602–613.
3. P. FrankPai, Bao-AnhNguyen, Mannur J. Sundaresan. Nonlinearity identification by time-domain-only signal processing, International Journal of Non-LinearMechanics, 54, (2013), 85–98.
4. H.-C. Chung, J. Liang, S. Kushiyama, M. Shinozuk, Digital image processing for non-linear system identification, International Journal of Non-Linear Mechanics, 39, (2004), 691 – 707.
5. Jing Na, Xuemei Ren, Yuanqing Xia, Adaptive parameter identification of linear SISO systems with unknown time-delay, Systems & Control Letters, 66, (2014), 43–50.
6. Osman Kukrer, Analysis of the dynamics of a memory less nonlinear gradient IIR adaptive notch filter, Signal Processing, 91(10), (2011), 2379–2394.
7. Tayebeh Mostajabi, Javad Poshtan, Zahra Mostajabi, IIR model identification via evolutionary algorithms, A comparative study, Artif Intell Rev, doi:10.1007/s10462-013-9403-1.
8. Dai, C., Chen, W., Zhu, Y., Seeker optimization algorithm for digital IIR filter design. IEEE Trans. Industr. Electron. 57 (5), (2010), 1710–1718.
9. Fang, W., Sun, J., Xu, W., A new mutated quantum behaved particle swarm optimizer for digital IIR filter. EURASIP J. Adv. Signal Process., (2009), article ID. 367465, 1–7.
10. J. Kennedy, R.C. Eberhart, Particle swarm optimization, in: Proceedings of the 1995 IEEE International Conference on Neural Networks, vol. 4, 1995, pp. 1942–1948.

11. D. Karaboga, An idea based on honey bee swarm for numerical optimization, Technical report,-TR06, Erciyes University, Engineering Faculty, Computer Engineering Department, 2005.
12. B. İlker, S. Birbil, F. Shu-Cherng, An Electromagnetism-like Mechanism for Global Optimization. Journal of Global Optimization, 25 (2003) 263–282.
13. X.-S. Yang, S. Deb, Cuckoo search via levy flights, in: World Congress on Nature Biologicall y Inspired Computing, 2009, pp. 210–214.
14. Yang, X. S. (2012), Flower pollination algorithm for global optimization, in: Unconventional Computation and Natural Computation, Lecture Notes in Computer Science, Vol. 7445, pp. 240–249.
15. Ahn, C., 2006. Advances in Evolutionary Algorithms: Theory, Design and Practice. Springer Publishing, New York.
16. Chiong, R., Weise, T., Michalewicz, Z., 2012. Variants of Evolutionary Algorithms for Real-World Applications. Springer, New York.
17. Oltean, M., 2007. Evolving evolutionary algorithms with patterns. Soft Comput. 11 (6), 503–518.
18. Chen, S., Luk, B.L., Digital IIR filter design using particle swarm optimization. Int. J. Model. Ident. Control 9 (4), (2010), 327–335.
19. Karaboga, N., A new design method based on artificial bee colony algorithm for digital IIR filters. J. Franklin Inst. 346 (4), (2009), 328–348.
20. Cuevas E., Oliva D., IIR Filter Modeling Using an Algorithm Inspired on Electromagnetism, Ingeniería Investigación y Tecnología, 14 (1), (2013), 125–138.
21. Apoorv P. Patwardhan, Rohan Patidar, Nithin V. George, On a cuckoo search optimization approach towards feedback system identification.
22. Wolpert, D.H., Macready, W.G., No Free Lunch Theorems for Optimization, IEEE Transactions on Evolutionary Computation 1(67), (1997), 67–82.
23. Emad Elbeltagi, Tarek Hegazy, Donald Grierson, Comparison among five evolutionary-based optimization algorithms, Advanced Engineering Informatics, 19, (2005), 43–53.
24. David Shilane, Jarno Martikainen, Sandrine Dudoit, Seppo J. Ovaska, A general framework for statistical performance comparison of evolutionary computation algorithms, Information Sciences 178, (2008), 2870–2879.
25. Valentın Osuna-Enciso, Erik Cuevas, Humberto Sossa, A comparison of nature inspired algorithms for multi-threshold image segmentation, Expert Systems with Applications, 40, (2013), 1213–1219.
26. Yih-Lon Lin, Wei-Der Chang, Jer-Guang Hsieh, A particle swarm optimization approach to nonlinear rational filter modeling, Expert Systems with Applications 34 (2008) 1194–1199.
27. Erik Cuevas, Mauricio González, Daniel Zaldivar, Marco Pérez-Cisneros, and Guillermo García, An Algorithm for Global Optimization Inspired by Collective Animal Behavior, Discrete Dynamics in Nature and Society, 2012 (2012), Article ID 638275, 24 pages.
28. Erik Cuevas, Miguel Cienfuegos, Daniel Zaldívar, Marco Pérez-Cisneros, A swarm optimization algorithm inspired in the behavior of the social-spider, Expert Systems with Applications 40 (2013) 6374–6384.
29. Oliva, D., Cuevas, E., Pajares, G., Zaldivar, D., Osuna, V., A Multilevel thresholding algorithm using electromagnetism optimization, Neurocomputing 139, (2014), 357–381.
30. Oliva, D., Cuevas, E., Pajares, G., Zaldivar, D., Perez-Cisneros, M., Multilevel thresholding segmentation based on harmony search optimization, Journal of Applied Mathematics, 2013, 575414.
31. Cuevas, E., Zaldivar, D., Pérez-Cisneros, M., Seeking multi-thresholds for image segmentation with Learning Automata, Machine Vision and Applications, 22 (5), (2011), 805–818.
32. Cuevas, E., Ortega-Sánchez, N., Zaldivar, D., Pérez-Cisneros, M., Circle detection by Harmony Search Optimization, Journal of Intelligent and Robotic Systems: Theory and Applications, 66 (3), (2012), 359–376.

33. Cuevas, E., Zaldivar, D., Pérez-Cisneros, M., Ramírez-Ortegón, M., Circle detection using discrete differential evolution Optimization, Pattern Analysis and Applications, 14 (1), (2011), 93–107.
34. Cuevas, E., Echavarría, A., Zaldívar, D., Pérez-Cisneros, M., A novel evolutionary algorithm inspired by the states of matter for template matching, Expert Systems with Applications, 40 (16), (2013), 6359–6373.
35. Garcia S, Molina D, Lozano M, Herrera F (2008) A study on the use of non-parametric tests for analyzing the evolutionary algorithms' behaviour: a case study on the CEC'2005 Special session on real parameter optimization. J Heurist. doi:10.1007/s10732-008-9080-4.
36. D. Shilane, J. Martikainen, S. Dudoit, S.. Ovaska. A general framework for statistical performance comparison of evolutionary computation algorithms. Information Sciences 178 (2008) 2870–2879.

Printed in the United States
By Bookmasters